Deep Inference and Symmetry in Classical Proofs

Dissertation zur Erlangung des akademischen Grades
Doktor rerum naturalium

vorgelegt an der
Technischen Universität Dresden
Fakultät Informatik

eingereicht von
Dipl.-Inf. Kai Brünnler
geboren am 28. Mai 1975 in Karl-Marx-Stadt

Gutachter:

Prof. Dr. rer. nat. habil. Steffen Hölldobler, Technische Universität Dresden
Prof. Dr. rer. nat. habil. Horst Reichel, Technische Universität Dresden
Prof. Dr. Dale Miller, INRIA Futurs und École Polytechnique

Tag der Verteidigung: 22. September 2003

Revised Version. Bern, December 2003.

Bibliografische Information Der Deutschen Bibliothek

Die Deutsche Bibliothek verzeichnet diese Publikation in der Deutschen
Nationalbibliografie; detaillierte bibliografische Daten sind im Internet über
http://dnb.ddb.de abrufbar.

ISBN 3-8325-0448-6

Logos Verlag Berlin
Comeniushof, Gubener Str. 47,
10243 Berlin
Tel.: +49 030 42 85 10 90
Fax: +49 030 42 85 10 92
INTERNET: http://www.logos-verlag.de

Abstract

In this thesis we see deductive systems for classical propositional and predicate logic which use *deep inference*, i.e. inference rules apply arbitrarily deep inside formulas, and a certain *symmetry*, which provides an involution on derivations. Like sequent systems, they have a cut rule which is admissible. Unlike sequent systems, they enjoy various new interesting properties. Not only the identity axiom, but also cut, weakening and even contraction are reducible to atomic form. This leads to inference rules that are *local*, meaning that the effort of applying them is bounded, and *finitely generating*, meaning that, given a conclusion, there is only a finite number of premises to choose from. The systems also enjoy new normal forms for derivations and, in the propositional case, a cut elimination procedure that is drastically simpler than the ones for sequent systems.

Acknowledgements

Alessio Guglielmi introduced me to proof theory. He deeply influenced my thoughts on the subject. His advice and support during the last years have been invaluable.

I also benefited from discussions with Paola Bruscoli, Steffen Hölldobler, Claus Jürgensen, Ozan Kahramanoğulları, Michel Parigot, Horst Reichel, Charles Stewart, Lutz Straßburger and Alwen Fernanto Tiu.

Außerdem danke ich Axel, Sebastian, Anni, Uli und Carsten, sowie Anton, Sylvia, Thomas, Glen, Lola, Nils, Heidi, Ronald, Micha, Mischa, Katya, Ricarda, Anne, Selina, Matti, Nicole, Jasper, Sheila, Holger, Juve und Roland. Ihr hattet alle auf die eine oder andere Weise einen positiven Einfluß auf diese Arbeit.

This work has been accomplished while I was supported by the DFG Graduiertenkolleg 'Spezifikation diskreter Prozesse und Prozeßsysteme durch operationelle Modelle und Logiken'.

Special thanks go to Paola Bruscoli for the picture on the cover, to Luigi Galvani, and to his dissected frogs. Poor guys.

Contents

Chapter 1

Introduction

The central idea of proof theory is cut elimination. Let us take a close look at the cut rule in the sequent calculus [16], in its one-sided version [34, 44]:

$$\text{Cut} \, \frac{\vdash \Phi, A \quad \vdash \Psi, \bar{A}}{\vdash \Phi, \Psi} \quad .$$

When seen bottom-up, it introduces an arbitrary formula A together with its negation $\bar{A}$. Another rule that introduces an arbitrary formula A and its negation $\bar{A}$, but this time when seen top-down, is the identity axiom

$$\text{Ax} \, \frac{}{\vdash A, \bar{A}} \quad .$$

Clearly, the two rules are intimately related. However, their duality is obscured by the fact that a certain top-down symmetry is inherently broken in the sequent calculus: derivations are trees, and trees are top-down asymmetric.

To reveal the duality between the two rules, let us now restore this symmetry. The tree-shape of derivations in the sequent calculus is of course due to the presence of two-premise inference rules. Consider for example the R∧-rule

$$\text{R}\wedge \, \frac{\vdash \Phi, A \quad \vdash \Psi, B}{\vdash \Phi, \Psi, A \wedge B} \quad ,$$

in which there is an asymmetry between premise and conclusion: two premises, but just one conclusion. Or, to put it differently: one connective in the conclusion, no connective between the premises.

This asymmetry can be repaired. We know that the comma in a sequent corresponds to disjunction and that the branching in the tree corresponds to conjunction, so we write the above rule as follows:

$$\frac{\vdash (\Phi \vee A) \wedge (\Psi \vee B)}{\vdash \Phi \vee \Psi \vee (A \wedge B)} \quad,$$

thereby identifying the 'logical' or 'object' level (the connectives in the formula) with the 'structural' or 'meta' level (the comma in a sequent and the branching of the proof tree).

Identifying the two levels of the sequent calculus in this way would render the system incomplete because one purpose of the tree-shape of derivations is to allow the inference rules to access subformulas. Consider the derivation

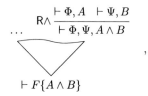

where the endsequent contains a formula that has a proper subformula $A \wedge B$. Seen bottom-up, the reason why the R$\wedge$-rule can eventually be applied to $A \wedge B$ is that the rules in the derivation below decompose the context $F\{\ \}$ and distribute its contents among the leaves of the derivation. Having dropped the tree-shape and the distinction between logical and structural level, we need to restore the ability of accessing subformulas. This can be done directly: we use *deep inference*, meaning that we allow inference rules to apply anywhere deep inside a formula, not only at the main connective. As usual in the one-sided sequent calculus, negation only occurs on atoms, so this is sound because implication is closed under disjunction, conjunction and quantification.

The identity axiom and the cut rule can now take the following form:

$$identity \; \frac{F\{true\}}{F\{A \vee \bar{A}\}} \qquad\qquad cut \; \frac{F\{A \wedge \bar{A}\}}{F\{false\}} \quad,$$

which makes precise the duality between the two: one can be obtained from the other by exchanging premise and conclusion and negating them. We see that this is the notion of duality well known under the name *contrapositive*.

We can now observe *symmetry*, meaning that all inference rules come in pairs of two dual rules, like identity and cut. This duality of inference rules extends to derivations: a derivation is dualised by negating everything and flipping it upside-down.

The *calculus of structures* by Guglielmi [22, 19] is a formalism which employs deep inference and symmetry. In contrast to the sequent calculus it allows us to observe the duality between cut and identity. The original purpose of this formalism was to express a logical system with a self-dual non-commutative connective resembling sequential composition in process algebras [22, 24, 25, 9]. Such a system cannot be expressed in the sequent calculus, as was shown by Tiu [43]. Then the calculus of structures turned out to yield systems with interesting properties also for existing logics. It has been employed by Straßburger to solve the problem of the non-local behaviour of the promotion rule in linear logic [42] and to give a local system for full linear logic [41]. Stewart and Stouppa give a collection of systems for modal logics which are simpler and more systematic than the corresponding collection of sequent systems [39]. Di Gianantonio gives a system for Yetter's cyclic linear logic in [12] which avoids the cycling rule. Classical logic in the calculus of structures is the subject of this thesis.

The calculus of structures is not alone in using deep inference. In [35, 36] Schütte defines inference rules that work deeply inside a formula on 'positive and negative parts of a formula'. These notions are related to the notions of positive and negative subformula, but they are different. Schütte suggests to see them as a generalisation of Gentzen's notions of antecedent and succedent and notes the fact that this generalisation removes the need for explicitly adding structural rules. Another system that uses deep inference is one by Pym for the substructural logic BI [31].

Top-down symmetry, on the other hand, seems to be a unique feature of the calculus of structures. At first sight, proof nets [18] appear to have a similar symmetry: in particular the axiom link and the cut link are dual. However, the ways in which they are used are not symmetric: the underlying structure of a proof net is made by trees and thus asymmetric; there is no easy involution on proof nets as there is on derivations in the calculus of structures.

The formalism of the rules we have just considered, the one-sided or *Gentzen-Schütte* sequent calculus, is symmetric in the sense that the De Morgan duality is built into the formulas in order to save half of the inference rules. For that reason it is also called *Gentzen-symmetric*. One could say that the symmetry in the calculus of structures is related, but it reaches much further: the De Morgan duality naturally extends from connectives to inference rules and thus from formulas to derivations. In the sequent calculus, no matter whether Gentzen-symmetric or not, this symmetry is inherently broken: derivations are trees, and trees are asymmetric.

In this thesis we will see that symmetry and deep inference, which so far have only been motivated by aesthetic considerations, allow us to develop a proof theory for classical logic much in the same way as in the sequent calculus and also yield new proof-theoretical properties that cannot be observed in the sequent calculus.

Many of the motivations for this work come from computer science. In fact, there is a fruitful relationship between proof theory and computer science. The most obvious link between the two is in the field of mechanised theorem proving: the common interest in formalising logical statements, inferences and proofs. However, the reasons for this interest differ. While in mechanised theorem proving the interest is mainly in efficient algorithms for finding proofs, proof theory studies the properties of proofs in a much more general way. Still, fundamental proof-theoretic results like Herbrand's Theorem [26] and related work have been foundational for mechanised theorem proving.

More interesting (but less obvious) links between proof theory and computer science are in the field of declarative programming languages. Central proof theoretic methods are connected to functional programming and to logic programming. In the paradigm of proof normalisation as computation, proofs correspond to functional programs and the normalisation of proofs corresponds to the evaluation of functional programs. This paradigm and its relevance for computer science is explained in [46]. In the paradigm of proof search as computation, searching for a proof corresponds to the execution of a logic program and a proof thus corresponds to the trace of a successful execution. The prime example of a logic programming language is Prolog [27].

By the two paradigms mentioned, properties of proofs correspond to properties of functional programs and to properties of logic programs. Of course, the correspondence of notions from proof theory to notions studied in computer science is by itself not so interesting to the computer scientist. It becomes interesting when techniques and results from proof theory can be transferred via this correspondence and can be applied in computer science. An example of the application of proof theoretic techniques in the realm of proof normalisation as computation is the proof of strong normalisation of the polymorphic λ-calculus by Girard [17], which is foundational for typed functional programming languages. An example of the application of proof theoretic results in the realm of proof search as normalisation is the programming language λ-Prolog [29, 28], which has several advantages over Prolog.

Summary of Results

There are four main results in this thesis:

- **Deep symmetric systems for classical logic.** In Chapter 2 and Chapter 3 we will see systems for propositional and predicate logic in the calculus of structures. Soundness and completeness are shown by translation to the sequent calculus. The inference rules of these systems have a finer granularity than the ones of the sequent calculus and the way of composing them is more general than in the sequent calculus. For these reasons, cut admissibility for these systems immediately follows from the given translations and cut admissibility of the sequent systems. The fact that this crucial property is not lost in moving from sequent systems to systems in the calculus of structures suggests to develop a proof theory in the calculus of structures in the same way as in sequent systems.

- **Locality and finite choice.** By easy modification of these systems we then obtain, in Chapter 4 and Chapter 5, systems whose inference rules are:

  - *local*, meaning that the application of an inference rule only affects a bounded portion of the formula it is applied to, and

  - *finitely generating*, meaning that, given the conclusion of an inference rule, there is only a finite number of premises to choose from.

  Locality implies a bounded computational cost of applying an inference rule and seems useful for distributed implementation. Locality is provably impossible in the sequent calculus. Finite choice can of course be achieved in the sequent calculus—through cut admissibility. In the calculus of structures it can be achieved with much simpler means.

- **Cut elimination.** The symmetry of the calculus of structures allows to reduce the cut to atomic form without having to go through cut elimination. In Chapter 6 we will see that this in turn allows for a very simple cut elimination procedure for propositional logic. In contrast to cut elimination procedures for the sequent calculus it involves neither permuting up the cut rule nor induction on the cut-rank. The atomicity of the cut rule allows for plugging proofs, similarly to what one can do in natural deduction. This suggests a computational interpretation along the lines of natural deduction.

- **Normal forms.** In Chapter 7 we will see new normal forms for derivations which divide derivations into distinct phases in a natural way. In some of them, these phases use rules from disjoint subsystems, which provides certain kinds of modularity. Most of these normal forms are provably impossible in the sequent calculus. They are available in the calculus of structures because there is more freedom in applying inference rules and consequently there are more permutations of rules to be observed. For the propositional case, one of the normal forms stands out in generalising both: cut elimination and Craig interpolation.

These results sustain the claim that the concepts of deep inference and top-down symmetry allow for a more refined combinatorial analysis of proofs than what is possible in the sequent calculus, while at the same time retaining the good properties of the sequent calculus, in particular cut admissibility.

Many results presented in this thesis have already appeared elsewhere, the local system for propositional logic in [8], the local system for predicate logic and the normal forms (but not interpolation) in [3], the impossibility of certain normal forms in the sequent calculus in [6], the finitely generating system in [7] and the cut elimination procedure in [4].

This thesis can be read as shown in Figure 1.1. Below the title of a chapter are the names of the systems introduced in this chapter. Readers who are just interested in propositional logic can skip Chapter 3 and the subsequent sections on predicate logic.

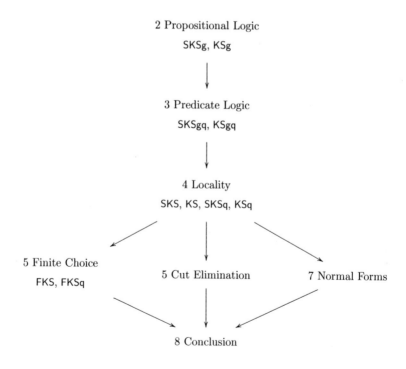

Figure 1.1: Chapter overview

Chapter 2

Propositional Logic

In this chapter we see a deductive system for classical propositional logic which follows the tradition of sequent systems, in particular there is a cut rule and its admissibility is shown. In contrast to sequent systems, its rules apply deep inside formulas and there is no branching in derivations. This allows to observe a vertical symmetry that can not be observed in the sequent calculus.

The chapter is structured as follows: after some basic definitions I present system SKSg, a set of inference rules for classical propositional logic which is closed under a notion of duality. I then translate derivations of a Gentzen-Schütte sequent system into this system, and vice versa. This establishes soundness and completeness with respect to classical propositional logic as well as cut admissibility.

2.1 Basic Definitions

Definition 2.1.1. *Propositional variables* v and their negations $\bar{v}$ are *atoms*. Atoms are denoted by a, b, ... The *formulas* of the language KS are generated by

$$S ::= \mathsf{f} \mid \mathsf{t} \mid a \mid [\underbrace{S, \ldots, S}_{>0}] \mid (\underbrace{S, \ldots, S}_{>0}) \mid \bar{S} \quad ,$$

where f and t are the units *false* and *true*, $[S_1, \ldots, S_h]$ is a *disjunction* and $(S_1, \ldots, S_h)$ is a *conjunction*. $\bar{S}$ is the *negation* of the formula S. Formulas are denoted by S, P, Q, R, T, U, V and W. *Formula contexts*, denoted by $S\{\ \}$, are formulas with one occurrence of $\{\ \}$, the *empty context* or *hole*, that does not appear in the scope of a negation. $S\{R\}$ denotes the formula obtained by filling the hole in $S\{\ \}$ with R. We drop the curly braces when they are redundant: for example, $S[R, T]$ is short for $S\{[R, T]\}$. A formula

7

Associativity   **Commutativity**

$$[\vec{R}, [\vec{T}], \vec{U}] = [\vec{R}, \vec{T}, \vec{U}]$$ $$[R, T] = [T, R]$$
$$(\vec{R}, (\vec{T}), \vec{U}) = (\vec{R}, \vec{T}, \vec{U})$$ $$(R, T) = (T, R)$$

Units   **Negation**

$$(\mathsf{f}, \mathsf{f}) = \mathsf{f} \qquad [\mathsf{f}, R] = R$$ $$\bar{\mathsf{f}} = \mathsf{t}$$
$$[\mathsf{t}, \mathsf{t}] = \mathsf{t} \qquad (\mathsf{t}, R) = R$$ $$\bar{\mathsf{t}} = \mathsf{f}$$

Context Closure $$\overline{[R, T]} = (\bar{R}, \bar{T})$$
$$\overline{(R, T)} = [\bar{R}, \bar{T}]$$

$$\text{if} \quad R = T \quad \text{then} \quad \frac{S\{R\} = S\{T\}}{\bar{R} = \bar{T}}$$ $$\bar{\bar{R}} = R$$

Figure 2.1: Syntactic equivalence of formulas

R is a *subformula* of a formula T if there is a context $S\{\ \}$ such that $S\{R\}$ is T. Formulas are (*syntactically*) *equivalent* modulo the smallest equivalence relation induced by the equations shown in Figure 2.1. There, $\vec{R}, \vec{T}$ and $\vec{U}$ are finite sequences of formulas, $\vec{T}$ is non-empty. Formulas are in *negation normal form* if negation occurs only on propositional variables.

For each formula there is an equivalent one in negation normal form. In the following, unless stated otherwise, I will assume formulas to be in negation normal form. Likewise, for each formula there is an equivalent one in which disjunction as well as conjunction only occur in their binary form. I will use this fact in induction arguments without explicitly mentioning it.

Example 2.1.2. The formulas $\overline{[a, b, c]}$ and $(\bar{a}, (\bar{b}, \bar{c}))$ are equivalent: the first is not in negation normal form, the second instead is. Contrary to the first, in the second formula disjunction and conjunction only occur in their binary form.

Definition 2.1.3. The letters denoting formulas, i.e. S, P, Q, are *schematic formulas*. Likewise, $S\{\ \}$ is a *schematic context*. An *inference rule* is a scheme written

$$\rho \frac{V}{U} \quad ,$$

where ρ is the *name* of the rule, V is its *premise* and U is its *conclusion*. Both U and V are formulas that each may contain schematic formulas and schematic contexts. If neither U nor V contain a schematic context, then the inference rule is called *shallow*, otherwise it is called *deep*. An *instance* of an inference rule is the inference rule in which each schematic context is replaced by a context and each schematic formula is replaced by a formula. If a deep inference rule is of the shape

$$\pi \frac{S\{T\}}{S\{R\}} \quad ,$$

where $S\{\ \}$ is a schematic context and R and T are formulas that may contain schematic formulas and contexts, then in an instance of π the formula taking the place of R is its *redex*, the formula taking the place of T is its *contractum* and the context taking the place of $S\{\ \}$ is its *context*. A (*deductive*) *system* $\mathscr{S}$ is a set of inference rules.

Most inference rules that we will consider have the shape of the rule π from the previous definition. Such an inference rule can be seen as a rewrite rule with the context made explicit. The rule π seen top-down corresponds to a rewrite rule $T \to R$. A shallow inference rule can be seen as a rewrite rule that may only be applied to the whole given term, not to arbitrary subterms.

Definition 2.1.4. A *derivation* Δ in a certain deductive system is a finite sequence of instances of inference rules in the system:

$$\pi \frac{T}{\underset{\pi'}{\overset{V}{}}} \\ \rho' \frac{\vdots}{\underset{\rho}{\overset{U}{}} } \frac{}{R} \ .$$

A derivation can consist of just one formula. The topmost formula in a derivation is called the *premise* of the derivation, and the formula at the bottom is called its *conclusion*.

Note that the notion of derivation is top-down symmetric, contrary to the corresponding notion in the sequent calculus.

Notation 2.1.5. A derivation Δ whose premise is T, whose conclusion is R, and whose inference rules are in $\mathscr{S}$ is denoted by

$$\underset{R}{\overset{T}{\Delta \| \mathscr{S}}} \ .$$

Deep inference allows to put derivations into a context to obtain a new derivation. This is related to the method of adding formulas to the context of every rule instance in a sequent calculus derivation.

Definition 2.1.6. Given a derivation Δ, the derivation $S\{\Delta\}$ is obtained as follows:

$$
\Delta = \quad
\begin{array}{c}
\pi' \dfrac{T}{V} \\
\pi \, \vdots \\
\rho' \dfrac{}{U} \\
\rho \dfrac{}{R}
\end{array}
\qquad
S\{\Delta\} = \quad
\begin{array}{c}
\pi' \dfrac{S\{T\}}{S\{V\}} \\
\pi \, \vdots \\
\rho' \dfrac{}{S\{U\}} \\
\rho \dfrac{}{S\{R\}}
\end{array}
\quad .
$$

Definition 2.1.7. The *equivalence rule* is

$$
= \frac{T}{R} \quad ,
$$

where R and T are syntactically equivalent formulas. Every system implicitly includes the equivalence rule. I will omit obvious instances of the equivalence rule from derivations.

Remark 2.1.8. Other works on systems in the calculus of structures define *structures* to be equivalence classes of formulas and inference rules that work on structures instead of formulas. The use of the term 'structure' for syntactic 'sequent-like' objects is common, cf. [1, 32]. However, this term is rather generic and could clash with the semantic notion of 'structure' as used in model theory. So instead of structures I use formulas and an equivalence rule. Of course one can trivially move from one approach to the other.

Definition 2.1.9. A rule ρ is *derivable* for a system $\mathscr{S}$ if for every instance of $\rho \dfrac{T}{R}$ there is a derivation $\begin{array}{c} T \\ \| \mathscr{S} \\ R \end{array}$.

The symmetry of derivations, where both premise and conclusion are arbitrary formulas, is broken in the notion of *proof*:

Definition 2.1.10. A *proof* is a derivation whose premise is the unit t. A proof Π of R in system $\mathscr{S}$ is denoted by

$$
\begin{array}{c}
\Pi \, \| \mathscr{S} \\
R
\end{array} \quad .
$$

2.2 A Deep Symmetric System

In this section we see a system for classical propositional logic which shows the following notion of duality, which is known as *contrapositive*:

Definition 2.2.1. The *dual* of an inference rule is obtained by exchanging premise and conclusion and replacing each connective by its De Morgan dual. For example

$$\mathsf{i}{\downarrow}\,\frac{S\{\mathsf{t}\}}{S[R,\bar{R}]} \qquad \text{is dual to} \qquad \mathsf{i}{\uparrow}\,\frac{S(R,\bar{R})}{S\{\mathsf{f}\}} \quad,$$

where the rule i↓ is called *identity* and the rule i↑ is called *cut*.

The rules i↓ and i↑ respectively correspond to the identity axiom and the cut rule in the sequent calculus, as we will see shortly.

Definition 2.2.2. A system of inference rules is called *symmetric* if for each of its rules it also contains the dual rule.

The symmetric system for propositional classical logic is shown in Figure 2.2. It is called system SKSg, where the first S stands for 'symmetric', K stands for 'klassisch' as in Gentzen's LK and the second S says that it is a system in the calculus of structures. Small letters are appended to the name of a system to denote variants. In this case, the g stands for 'general', meaning that rules are not restricted to atoms: they can be applied to arbitrary formulas. We will see in the next section that this system is sound and complete for classical propositional logic.

The rules s, w↓ and c↓ are called respectively *switch*, *weakening* and *contraction*. Their dual rules carry the same name prefixed with a 'co-', so e.g. w↑ is called *co-weakening*. Rules i↓, w↓, c↓ are called *down-rules* and their duals are called *up-rules*. The dual of the switch rule is the switch rule itself: it is *self-dual*.

I now try to give an idea on how the familiar rules of the sequent calculus correspond to the rules of SKSg. For the sake of simplicity I consider the rules of the sequent calculus in isolation, i.e. not as part of a proof tree. The full correspondence is shown in Section 2.3.

The identity axiom of the sequent calculus corresponds to the identity rule i↓:

$$\overline{\vdash A,\bar{A}} \qquad \text{corresponds to} \qquad \mathsf{i}{\downarrow}\,\frac{\mathsf{t}}{[A,\bar{A}]} \quad.$$

$$\text{i}{\downarrow}\ \frac{S\{\text{t}\}}{S[R,\bar{R}]} \qquad\qquad \text{i}{\uparrow}\ \frac{S(R,\bar{R})}{S\{\text{f}\}}$$

$$\text{s}\ \frac{S([R,U],T)}{S[(R,T),U]}$$

$$\text{w}{\downarrow}\ \frac{S\{\text{f}\}}{S\{R\}} \qquad\qquad \text{w}{\uparrow}\ \frac{S\{R\}}{S\{\text{t}\}}$$

$$\text{c}{\downarrow}\ \frac{S[R,R]}{S\{R\}} \qquad\qquad \text{c}{\uparrow}\ \frac{S\{R\}}{S(R,R)}$$

Figure 2.2: System SKSg

However, i↓ can appear anywhere in a proof, not only at the top. The cut rule of the sequent calculus corresponds to the rule i↑ followed by two instances of the switch rule:

$$\text{Cut}\ \frac{\vdash \Phi,A \quad \vdash \Psi,\bar{A}}{\vdash \Phi,\Psi} \qquad \text{corresponds to} \qquad \begin{array}{l} \text{s}\ \dfrac{([\Phi,A],[\Psi,\bar{A}])}{} \\[4pt] \text{s}\ \dfrac{[\Phi,(A,[\Psi,\bar{A}])]}{} \\[4pt] \text{i}{\uparrow}\ \dfrac{[\Phi,\Psi,(A,\bar{A})]}{} \\[4pt] =\ \dfrac{[\Phi,\Psi,\text{f}]}{[\Phi,\Psi]} \end{array} \quad .$$

The multiplicative (or context-splitting) R∧ rule in the sequent calculus corresponds to two instances of the switch rule:

$$\text{R}{\wedge}\ \frac{\vdash \Phi,A \quad \vdash \Psi,B}{\vdash \Phi,\Psi,A\wedge B} \qquad \text{corresponds to} \qquad \begin{array}{l} \text{s}\ \dfrac{([\Phi,A],[\Psi,B])}{} \\[4pt] \text{s}\ \dfrac{[\Phi,(A,[\Psi,B])]}{[\Phi,\Psi,(A,B)]} \end{array} \quad .$$

A contraction in the sequent calculus corresponds to the c↓ rule:

$$\text{RC}\ \frac{\vdash \Phi,A,A}{\vdash \Phi,A} \qquad \text{corresponds to} \qquad \text{c}{\downarrow}\ \frac{[\Phi,A,A]}{[\Phi,A]} \quad ,$$

just as the weakening in the sequent calculus corresponds to the w↓ rule:

$$\mathsf{RW}\,\frac{\vdash \Phi}{\vdash \Phi, A} \qquad \text{corresponds to} \qquad \mathsf{w}\!\downarrow\frac{\stackrel{\Phi}{=}}{[\Phi, A]}$$

$$\mathsf{w}\!\downarrow\frac{\frac{\Phi}{[\Phi, \mathsf{f}]}}{[\Phi, A]} \quad .$$

The c↑ and w↑ rules have no analogue in the sequent calculus. Their role is to ensure that our system is symmetric. They are obviously sound since they are just duals of the rules c↓ and w↓ which correspond to sequent calculus rules.

Derivations in a symmetric system can be dualised:

Definition 2.2.3. The *dual* of a derivation is obtained by turning it upside-down and replacing each rule, each connective and each atom by its dual. For example

$$\mathsf{c}\!\downarrow\frac{\mathsf{w}\!\uparrow\frac{[(a, \bar{b}), a]}{[a, a]}}{a} \qquad \text{is dual to} \qquad \mathsf{w}\!\downarrow\frac{\mathsf{c}\!\uparrow\frac{\bar{a}}{(\bar{a}, \bar{a})}}{([\bar{a}, b], \bar{a})} \quad .$$

As we will see in the next section, a formula T implies a formula R if and only if there is a derivation from T to R. So derivations correspond to implications. Dualising a derivation from T to R, illustrated in Figure 2.3, yields a derivation from $\bar{R}$ to $\bar{T}$, and vice versa. On the corresponding implications this duality is known as contraposition.

Figure 2.3: Symmetry

The notion of proof, however, is an asymmetric one: the dual of a proof is not a proof. Instead it is a derivation whose conclusion is the unit f. This dual notion of proof will be called *refutation*.

We now see that one can easily move back and forth between a derivation and a proof of the corresponding implication via the deduction theorem:

Theorem 2.2.4 (Deduction Theorem).

There is a derivation $\begin{array}{c} T \\ \Big\Vert \mathsf{SKSg} \\ R \end{array}$ *if and only if there is a proof* $\begin{array}{c} \Big\Vert \mathsf{SKSg} \\ [\bar{T}, R] \end{array}$.

Proof. A proof Π can be obtained from a given derivation Δ as follows:

$$\Delta \begin{array}{c} T \\ \Big\Vert \mathsf{SKSg} \\ R \end{array} \quad \rightsquigarrow \quad \mathsf{i}{\downarrow} \frac{\mathsf{t}}{[\bar{T}, T]} \atop [\bar{T}, \Delta] \Big\Vert \mathsf{SKSg} \atop [\bar{T}, R]} \quad ,$$

and a derivation Δ from a given proof Π as follows:

$$\Pi \begin{array}{c} \Big\Vert \mathsf{SKSg} \\ [\bar{T}, R] \end{array} \quad \rightsquigarrow \quad (T, \Pi) \Big\Vert \mathsf{SKSg} \atop \mathsf{s} \frac{(T, [\bar{T}, R])}{\mathsf{i}{\uparrow} \frac{[R, (T, \bar{T})]}{R}} \quad .$$

$\square$

2.3   Correspondence to the Sequent Calculus

The sequent system that is most similar to system SKSg is the one-sided system GS1p [44], also called *Gentzen-Schütte* system. In this section we consider a version of GS1p with multiplicative context treatment and constants $\top$ and $\bot$, and we translate its derivations to derivations in SKSg and vice versa. Translating from the sequent calculus to the calculus of structures is straightforward, in particular, no new cuts are introduced in the process. To translate in the other direction we have to simulate deep inferences in the sequent calculus, which is done by using the cut rule.

One consequence of those translations is that system SKSg is sound and complete for classical propositional logic. Another consequence is cut elimination: one can translate a proof with cuts in SKSg to a proof in GS1p + Cut, apply cut elimination for GS1p, and translate back the resulting cut-free proof to obtain a cut-free proof in SKSg.

$$\top \; \frac{}{\vdash \top} \qquad\qquad \mathsf{Ax} \; \frac{}{\vdash A, \bar{A}}$$

$$\mathsf{R}\wedge \; \frac{\vdash \Phi, A \quad \vdash \Psi, B}{\vdash \Phi, \Psi, A \wedge B} \qquad\qquad \mathsf{R}\vee \; \frac{\vdash \Phi, A, B}{\vdash \Phi, A \vee B}$$

$$\mathsf{RC} \; \frac{\vdash \Phi, A, A}{\vdash \Phi, A} \qquad\qquad \mathsf{RW} \; \frac{\vdash \Phi}{\vdash \Phi, A}$$

Figure 2.4: GS1p: classical logic in Gentzen-Schütte form

Definition 2.3.1. System GS1p is the set of rules shown in Figure 2.4. The system GS1p + Cut is GS1p together with

$$\mathsf{Cut} \; \frac{\vdash \Phi, A \quad \vdash \Psi, \bar{A}}{\vdash \Phi, \Psi} \qquad .$$

Formulas of system GS1p are denoted by A and B. They contain negation only on atoms and may contain the constants $\top$ and $\bot$. Multisets of formulas are denoted by Φ and Ψ. The empty multiset is denoted by $\varnothing$. In $A_1, \ldots, A_h$, where $h \geq 0$, a formula denotes the corresponding singleton multiset and the comma denotes multiset union. *Sequents*, denoted by Σ, are multisets of formulas. *Derivations* of system GS1p are trees denoted by Δ or by

$$\underset{\Sigma}{\overset{\Sigma_1 \; \cdots \; \Sigma_h}{\bigtriangledown \Delta}} \quad ,$$

where $h \geq 0$, the sequents $\Sigma_1, \ldots, \Sigma_h$ are the *premises* and Σ is the *conclusion*. A leaf of a derivation is *closed* if is either an instance of Ax or an instance of $\top$. *Proofs*, denoted by Π, are derivations where each leaf is closed.

From Sequent Calculus to Calculus of Structures

Definition 2.3.2. The function $\underline{}_s$ maps formulas, multisets of formulas and sequents of GS1p to formulas of KS:

$$\underline{a}_s = a \,,$$

$$\underline{\top}_s = t \,,$$

$$\underline{\bot}_s = f \,,$$

$$\underline{A \vee B}_s = [\underline{A}_s, \underline{B}_s] \,,$$

$$\underline{A \wedge B}_s = (\underline{A}_s, \underline{B}_s) \,,$$

$$\underline{\varnothing}_s = f \,,$$

$$\underline{A_1, \ldots, A_h}_s = [\underline{A_1}_s, \ldots, \underline{A_h}_s] \,, \qquad \text{where } h > 0.$$

In proofs, when no confusion is possible, the subscript $\underline{}_s$ may be dropped to improve readability.

Theorem 2.3.3. *For every derivation* $\displaystyle \overset{\Sigma_1 \ \cdots \ \Sigma_h}{\underset{\Sigma}{\bigtriangledown}}$ *in* GS1p + Cut *there exists*

a derivation $\displaystyle \underset{\underline{\Sigma}_s}{\overset{(\underline{\Sigma_1}_s, \ldots, \underline{\Sigma_h}_s)}{\big\| \text{SKSg} \setminus \{c{\uparrow}, w{\uparrow}\}}}$ *with the same number of cuts.*

Proof. By structural induction on the given derivation Δ.

Base Cases

1. $\Delta = \Sigma$. Take $\underline{\Sigma}_s$.

2. $\Delta = \top \dfrac{}{\vdash \top}$. Take t .

3. $\Delta = \text{Ax} \dfrac{}{\vdash A, \bar{A}}$. Take $i{\downarrow} \dfrac{t}{[\underline{A}_s, \underline{\bar{A}}_s]}$.

Inductive Cases

In the case of the R$\wedge$ rule, we have a derivation

$$\Delta \quad = \quad \begin{array}{c} \Sigma_1 \cdots \Sigma_k \quad \Sigma_1' \cdots \Sigma_l' \\ \bigtriangledown \qquad \bigtriangledown \\ \mathsf{R}\wedge \dfrac{\vdash \Phi, A \qquad \vdash \Psi, B}{\vdash \Phi, \Psi, A \wedge B} \end{array} \quad .$$

By induction hypothesis we obtain derivations

$$\begin{array}{c} (\Sigma_1, \ldots, \Sigma_k) \\ \Delta_1 \Big\| \mathsf{SKSg} \setminus \{\mathsf{c}\!\uparrow, \mathsf{w}\!\uparrow\} \\ [\Phi, A] \end{array} \quad \text{and} \quad \begin{array}{c} (\Sigma_1', \ldots, \Sigma_l') \\ \Delta_2 \Big\| \mathsf{SKSg} \setminus \{\mathsf{c}\!\uparrow, \mathsf{w}\!\uparrow\} \\ [\Psi, B] \end{array} \quad .$$

The derivation Δ_1 is put into the context $(\{\ \}, \Sigma_1', \ldots, \Sigma_l')$ to obtain Δ_1' and the derivation Δ_2 is put into the context $([\Phi, A], \{\ \})$ to obtain Δ_2':

$$\begin{array}{c} (\Sigma_1, \ldots, \Sigma_k, \Sigma_1', \ldots, \Sigma_l') \\ \Delta_1' \Big\| \mathsf{SKSg} \setminus \{\mathsf{c}\!\uparrow, \mathsf{w}\!\uparrow\} \\ ([\Phi, A], \Sigma_1', \ldots, \Sigma_l') \end{array} \quad \text{and} \quad \begin{array}{c} ([\Phi, A], \Sigma_1', \ldots, \Sigma_l') \\ \Delta_2' \Big\| \mathsf{SKSg} \setminus \{\mathsf{c}\!\uparrow, \mathsf{w}\!\uparrow\} \\ ([\Phi, A], [\Psi, B]) \end{array} \quad .$$

The derivation in SKSg we are looking for is obtained by composing Δ_1' and Δ_2' and applying the switch rule twice:

$$\begin{array}{c} (\Sigma_1, \ldots, \Sigma_k, \Sigma_1', \ldots, \Sigma_l') \\ \Big\| \\ \Delta_1' \Big\| \mathsf{SKSg} \setminus \{\mathsf{c}\!\uparrow, \mathsf{w}\!\uparrow\} \\ \\ ([\Phi, A], \Sigma_1', \ldots, \Sigma_l') \\ \Big\| \\ \Delta_2' \Big\| \mathsf{SKSg} \setminus \{\mathsf{c}\!\uparrow, \mathsf{w}\!\uparrow\} \\ \\ \mathsf{s} \dfrac{([\Phi, A], [\Psi, B])}{\mathsf{s} \dfrac{[\Psi, ([\Phi, A], B)]}{[\Phi, \Psi, (A, B)]}} \end{array} \quad .$$

The other cases are similar. The only case that requires a cut in SKSg is a cut in $\mathsf{GS1p}$. $\qquad\square$

Since proofs are special derivations, we obtain the following corollaries:

Corollary 2.3.4. *If a sequent Σ has a proof in* $\mathsf{GS1p} + \mathsf{Cut}$ *then* $\underline{\Sigma}_s$ *has a proof in* $\mathsf{SKSg} \setminus \{\mathsf{c}\!\uparrow, \mathsf{w}\!\uparrow\}$.

Corollary 2.3.5. *If a sequent Σ has a proof in* $\mathsf{GS1p}$ *then* $\underline{\Sigma}_s$ *has a proof in* $\mathsf{SKSg} \setminus \{\mathsf{i}\!\uparrow, \mathsf{c}\!\uparrow, \mathsf{w}\!\uparrow\}$.

From Calculus of Structures to Sequent Calculus

In the following we assume that formula of the language KS contain negation only on atoms and only conjunctions and disjunctions of exactly two formulas.

Definition 2.3.6. The function $\underline{\;\cdot\;}_G$ maps formulas of SKSg to formulas of GS1p:

$$
\begin{array}{rcl}
\underline{a}_G & = & a, \\
\underline{t}_G & = & \top, \\
\underline{f}_G & = & \bot, \\
\underline{[R,T]}_G & = & \underline{R}_G \vee \underline{T}_G, \\
\underline{(R,T)}_G & = & \underline{R}_G \wedge \underline{T}_G.
\end{array}
$$

We need the following lemma to imitate deep inference in the sequent calculus:

Lemma 2.3.7. *For every two formulas A, B and every formula context $C\{\ \}$*

there exists a derivation

$$
\begin{array}{c}
\vdash A, \bar{B} \\
\diagdown\diagup \\
\vdash C\{A\}, \overline{C\{B\}}
\end{array}
$$

in GS1p.

Proof. By structural induction on the context $C\{\ \}$. The base case in which $C\{\ \} = \{\ \}$ is trivial. If $C\{\ \} = C_1 \wedge C_2\{\ \}$, then the derivation we are looking for is

$$
\cfrac{\mathrm{Ax}\ \cfrac{}{\vdash C_1, \bar{C}_1}\quad \mathrm{R}\wedge\ \cfrac{\begin{array}{c}\vdash A, \bar{B}\\ \diagdown\!\!\!{}_{\Delta}\!\!\!\diagup\\ \vdash C_2\{A\}, \overline{C_2\{B\}}\end{array}}{}}{\mathrm{R}\vee\ \cfrac{\vdash C_1 \wedge C_2\{A\}, \bar{C}_1, \overline{C_2\{B\}}}{\vdash C_1 \wedge C_2\{A\}, \bar{C}_1 \vee \overline{C_2\{B\}}}}\quad,
$$

where Δ exists by induction hypothesis. The other case, in which $C\{\ \} = C_1 \vee C_2\{\ \}$, is similar. $\square$

Now we can translate derivations in system SKSg into derivations in the sequent calculus:

Theorem 2.3.8. *For every derivation* $\left\| \begin{smallmatrix} Q \\ \text{SKSg} \\ P \end{smallmatrix} \right.$ *there exists a derivation*

$$\underbrace{\vdash Q}_{G} \atop \underbrace{\vdash P}_{G}$$

in GS1p + Cut.

Proof. We construct the sequent derivation by induction on the length of the given derivation Δ in SKSg.

Base Case

If Δ consists of just one formula P, then P and Q are the same. Take $\vdash \underline{P}_G$.

Inductive Cases

We single out the topmost rule instance in Δ:

$$\Delta \left\| \begin{smallmatrix} Q \\ \text{SKSg} \\ P \end{smallmatrix} \right. \quad = \quad \rho \dfrac{S\{T\}}{S\{R\}} \atop \Delta' \left\| \begin{smallmatrix} \\ \text{SKSg} \\ P \end{smallmatrix} \right.$$

The corresponding derivation in GS1p will be as follows:

$$\text{Cut} \dfrac{ \begin{array}{c} \overbrace{\hspace{2cm}}^{\Pi} \\ \vdash R, \bar{T} \\ \overbrace{\hspace{2cm}}^{\Delta_1} \\ \vdash S\{R\}, \overline{S\{T\}} \quad \vdash S\{T\} \end{array} }{ \vdash S\{R\} } $$
$$\begin{array}{c} \overbrace{\hspace{2cm}}^{\Delta_2} \\ \vdash P \end{array}$$

where Δ_1 exists by Lemma 2.3.7 and Δ_2 exists by induction hypothesis. The proof Π depends on the rule ρ. In the following we will see that the proof Π exists for all the rules of SKSg.

For identity and cut, i.e.

$$\mathsf{i}{\downarrow}\,\frac{S\{\mathsf{t}\}}{S[U,\bar{U}]} \qquad \text{and} \qquad \mathsf{i}{\uparrow}\,\frac{S(U,\bar{U})}{S\{\mathsf{f}\}} \quad ,$$

we have the following proofs:

$$\mathsf{RW}\,\frac{\mathsf{RV}\,\dfrac{\mathsf{Ax}\,\overline{\vdash U,\bar{U}}}{\vdash U\vee\bar{U}}}{\vdash U\vee\bar{U},\bot} \qquad \text{and} \qquad \mathsf{RW}\,\frac{\mathsf{RV}\,\dfrac{\mathsf{Ax}\,\overline{\vdash U,\bar{U}}}{\vdash \bar{U}\vee U}}{\vdash \bot,\bar{U}\vee U} \quad .$$

In the case of the switch rule, i.e.

$$\mathsf{s}\,\frac{S([U,V],T)}{S[(U,T),V]} \quad ,$$

we have

$$\mathsf{RV}^2\,\frac{\mathsf{RA}\,\dfrac{\mathsf{RA}\,\dfrac{\mathsf{Ax}\,\overline{\vdash U,\bar{U}}\quad \mathsf{Ax}\,\overline{\vdash V,\bar{V}}}{\vdash U,\bar{U}\wedge\bar{V},V}\quad \mathsf{Ax}\,\overline{\vdash T,\bar{T}}}{\vdash (U\wedge T),V,(\bar{U}\wedge\bar{V}),\bar{T}}}{\vdash (U\wedge T)\vee V,(\bar{U}\wedge\bar{V})\vee\bar{T}} \quad ,$$

where RV^2 denotes two instances of the RV rule.

For contraction and its dual, i.e.

$$\mathsf{c}{\downarrow}\,\frac{S[U,U]}{S\{U\}} \qquad \text{and} \qquad \mathsf{c}{\uparrow}\,\frac{S\{U\}}{S(U,U)} \quad ,$$

we have

$$\mathsf{RC}\,\frac{\mathsf{RA}\,\dfrac{\mathsf{Ax}\,\overline{\vdash U,\bar{U}}\quad \mathsf{Ax}\,\overline{\vdash U,\bar{U}}}{\vdash U,U,\bar{U}\wedge\bar{U}}}{\vdash U,\bar{U}\wedge\bar{U}} \qquad \text{and} \qquad \mathsf{RC}\,\frac{\mathsf{RA}\,\dfrac{\mathsf{Ax}\,\overline{\vdash U,\bar{U}}\quad \mathsf{Ax}\,\overline{\vdash U,\bar{U}}}{\vdash U\wedge U,\bar{U},\bar{U}}}{\vdash U\wedge U,\bar{U}} \quad .$$

For weakening and its dual, i.e.

$$\mathsf{w}{\downarrow}\,\frac{S\{\mathsf{f}\}}{S\{U\}} \qquad \text{and} \qquad \mathsf{w}{\uparrow}\,\frac{S\{U\}}{S\{\mathsf{t}\}} \quad ,$$

we have

$$\mathsf{RW}\,\frac{\top\,\dfrac{}{\vdash \top}}{\vdash U,\top} \qquad \text{and} \qquad \mathsf{RW}\,\frac{\top\,\dfrac{}{\vdash \top}}{\vdash \top,\bar{U}} \quad .$$

$\square$

Corollary 2.3.9. *If a formula S has a proof in* SKSg *then* $\vdash \underline{S}_G$ *has a proof in* GS1p + Cut.

Soundness and completeness of SKSg, i.e. the fact that a formula has a proof if and only if it is valid, follows from soundness and completeness of GS1p by Corollaries 2.3.4 and 2.3.9. By symmetry a formula has a refutation if and only if it is unsatisfiable. Moreover, a formula T implies a formula R if and only if there is a derivation from T to R, which follows from soundness and completeness and the deduction theorem.

2.4 Cut Admissibility

In this section we see that if one is just interested in provability, then the up-rules of the symmetric system SKSg, i.e. i↑, w↑ and c↑, are superfluous. By removing them we obtain the asymmetric, cut-free system shown in Figure 2.5, which is called system KSg.

Definition 2.4.1. A rule ρ is *admissible* for a system $\mathscr{S}$ if for every proof $\left\| {}^{\mathscr{S} \cup \{\rho\}}_{} \atop S \right.$ there is a proof $\left\| {}^{\mathscr{S}}_{} \atop S \right.$.

The admissibility of all the up-rules for system KSg is shown by using the translation functions from the previous section:

Theorem 2.4.2. *The rules* i↑, w↑ *and* c↑ *are admissible for system* KSg.

Proof.

$$\left\|_{S}^{\text{SKSg}} \xrightarrow{\text{Corollary 2.3.9}} \overset{\text{GS1p}}{\underset{+\text{Cut}}{\bigtriangledown}} \underset{\vdash \underline{S}_G}{} \xrightarrow{\substack{\text{Cut elimination} \\ \text{for GS1p}}} \overset{\text{GS1p}}{\underset{\vdash \underline{S}_G}{\bigtriangledown}} \xrightarrow{\text{Corollary 2.3.5}} \left\|_{S}^{\text{KSg}}\right.$$

□

Definition 2.4.3. Two systems $\mathscr{S}$ and $\mathscr{S}'$ are *(weakly) equivalent* if for every proof $\left\| {}^{\mathscr{S}}_{} \atop R \right.$ there is a proof $\left\| {}^{\mathscr{S}'}_{} \atop R \right.$, and vice versa.

Corollary 2.4.4. *The systems* SKSg *and* KSg *are equivalent.*

Definition 2.4.5. Two systems $\mathscr{S}$ and $\mathscr{S}'$ are *strongly equivalent* if for every derivation $\overset{T}{\underset{R}{\|}}_{\mathscr{S}}$ there is a derivation $\overset{T}{\underset{R}{\|}}_{\mathscr{S}'}$, and vice versa.

$$\mathsf{i}{\downarrow}\,\frac{S\{\mathsf{t}\}}{S[R,\bar R]}\qquad\qquad \mathsf{s}\,\frac{S([R,T],U)}{S[(R,U),T]}\qquad\qquad \mathsf{w}{\downarrow}\,\frac{S\{\mathsf{f}\}}{S\{R\}}\qquad\qquad \mathsf{c}{\downarrow}\,\frac{S[R,R]}{S\{R\}}$$

Figure 2.5: System KSg

Remark 2.4.6. The systems SKSg and KSg are not strongly equivalent. The cut rule, for example, can not be derived in system KSg.

When a formula R implies a formula T then there is not necessarily a derivation from R to T in KSg, while there is one in SKSg. Therefore, I will in general use the asymmetric, cut-free system for deriving conclusions from the unit t, while I will use the symmetric system (i.e. the system with cut) for deriving conclusions from arbitrary premises.

As a result of cut elimination, sequent systems fulfill the subformula property. Our case is different, because the notions of formula and sequent are merged. Technically, system KSg does not fulfill the subformula property, just as system GS1p does not fulfill a 'subsequent property'. However, seen bottom-up, in system KSg no rule introduces new atoms. It thus satisfies the main aspect of the subformula property: when given a conclusion of a rule there is only a finite number of premises to choose from. In proof search, for example, the branching of the search tree is finite.

There is also a semantic cut elimination proof for system SKSg, analogous to the one given in [44] for system G3. The given proof with cuts is thrown away, keeping only the information that its conclusion is valid, and a cut-free proof is constructed from scratch. This actually gives us more than just admissibility of the up-rules: it also yields a separation of proofs into distinct phases.

Definition 2.4.7. A rule ρ is *invertible* for a system $\mathscr{S}$ if for each instance $\rho\,\dfrac{V}{U}$ there is a derivation $\left.\parallel\right.\!\!_{\mathscr{S}}\begin{array}{c}U\\ \\ V\end{array}$.

Theorem 2.4.8.

For every proof $\left\|\right.\!^{\mathsf{SKSg}}_{\;S}$ there is a proof $\begin{array}{c}\parallel\{\mathsf{i}{\downarrow}\}\\ S''\\ \parallel\{\mathsf{w}{\downarrow}\}\\ S'\\ \parallel\{\mathsf{s},\mathsf{c}{\downarrow}\}\\ S\end{array}$.

Proof. Consider the rule *distribute*:

$$\mathsf{d} \, \frac{S([R,T],[R,U])}{S[R,(T,U)]} \quad,$$

which can be derived by a contraction and two switches:

$$\mathsf{c}{\downarrow} \, \frac{\mathsf{s} \, \frac{\mathsf{s} \, \frac{S([R,T],[R,U])}{S[R,([R,T],U)]}}{S[R,R,(T,U)]}}{S[R,(T,U)]} \quad.$$

Build a derivation $\begin{array}{c} S' \\ \|_{\{\mathsf{d}\}} \\ S \end{array}$, by going upwards from S applying d as many times as possible. Then S' will be in conjunctive normal form, i.e.

$$S' = ([a_{11}, a_{12}, \ldots], [a_{21}, a_{22}, \ldots], \ldots, [a_{n1}, a_{n2}, \ldots]) \quad.$$

S is valid because there is a proof of it. The rule d is invertible, so S' is also valid. A conjunction is valid only if all its immediate subformulas are valid. Those are disjunctions of atoms. A disjunction of atoms is valid only if it contains an atom a together with its negation $\bar{a}$. Thus, more specifically, S' is of the following form (where we disregard the order of atoms in a disjunction):

$$S' = ([b_1, \bar{b}_1, a_{11}, a_{12}, \ldots], [b_2, \bar{b}_2, a_{21}, a_{22}, \ldots], \ldots, [b_n, \bar{b}_n, a_{n1}, a_{n2}, \ldots]) \quad.$$

Let $S'' = ([b_1, \bar{b}_1], [b_2, \bar{b}_2], \ldots, [b_n, \bar{b}_n])$.

Obviously, there is a derivation $\begin{array}{c} S'' \\ \|_{\{\mathsf{w}{\downarrow}\}} \\ S' \end{array}$ and a proof $\begin{array}{c} \| \, {\{\mathsf{i}{\downarrow}\}} \\ S'' \end{array}$. $\quad\square$

Chapter Summary

We have seen a deductive system for classical propositional logic. Its derivations easily correspond to derivations in the one-sided sequent calculus, which grants cut admissibility. Similarly to system $\mathsf{G3}$ it admits a simple semantic cut admissibility argument.

In contrast to sequent systems, its rules apply deep inside formulas and there is no branching in derivations. This allowed us to observe a vertical symmetry, the duality of derivations which consists in negating and flipping. This symmetry can not be observed in the sequent calculus.

Chapter 3

Predicate Logic

In this chapter we see a system for predicate logic. The use of deep inference allows to design this system in such a way that each rule corresponds to an implication from premise to conclusion, which is not true in the sequent calculus. Also, the eigenvariable conditions in this system are local, in contrast to the sequent calculus, where checking the eigenvariable condition requires checking the entire context.

This chapter is structured as the previous one: after some basic definitions I present system SKSgq, a set of inference rules for classical predicate logic which is closed under a notion of duality. I then translate derivations of a Gentzen-Schütte sequent system into this system, and vice versa. This establishes soundness and completeness with respect to classical predicate logic as well as cut admissibility.

3.1 Basic Definitions

Definition 3.1.1. *Variables* are denoted by x and y. *Terms* are denoted by τ and are defined as usual in first-order predicate logic. *Atoms*, denoted by a, b, etc., are expressions of the form $p(\tau_1, \ldots, \tau_n)$, where p is a *predicate symbol* of *arity* n and $\tau_1, \ldots, \tau_n$ are terms. The negation of an atom is again an atom. The *formulas* of the language KSq are generated by the following grammar, which is the one for the propositional case extended by existential and universal quantifier:

$$S ::= \mathsf{f} \mid \mathsf{t} \mid a \mid [\underbrace{S, \ldots, S}_{>0}] \mid (\underbrace{S, \ldots, S}_{>0}) \mid \bar{S} \mid \exists x S \mid \forall x S \quad .$$

Definition 3.1.2. Formulas are *equivalent* modulo the smallest equivalence relation induced by the equations in Figure 2.1 extended by the following

25

$$\text{i}\downarrow \frac{S\{\text{t}\}}{S[R,\bar{R}]} \qquad\qquad \text{i}\uparrow \frac{S(R,\bar{R})}{S\{\text{f}\}}$$

$$\text{s}\ \frac{S([R,U],T)}{S[(R,T),U]}$$

$$\text{u}\downarrow \frac{S\{\forall x[R,T]\}}{S[\forall xR,\exists xT]} \qquad\qquad \text{u}\uparrow \frac{S(\exists xR,\forall xT)}{S\{\exists x(R,T)\}}$$

$$\text{w}\downarrow \frac{S\{\text{f}\}}{S\{R\}} \qquad\qquad \text{w}\uparrow \frac{S\{R\}}{S\{\text{t}\}}$$

$$\text{c}\downarrow \frac{S[R,R]}{S\{R\}} \qquad\qquad \text{c}\uparrow \frac{S\{R\}}{S(R,R)}$$

$$\text{n}\downarrow \frac{S\{R[x/\tau]\}}{S\{\exists xR\}} \qquad\qquad \text{n}\uparrow \frac{S\{\forall xR\}}{S\{R[x/\tau]\}}$$

Figure 3.1: System SKSgq

equations:

Variable Renaming $\begin{aligned}\forall xR &= \forall yR[x/y]\\ \exists xR &= \exists yR[x/y]\end{aligned}$ if y is not free in R

Vacuous Quantifier $\forall yR = \exists yR = R$ if y is not free in R

Negation $\begin{aligned}\overline{\exists xR} &= \forall x\bar{R}\\ \overline{\forall xR} &= \exists x\bar{R}\end{aligned}$

Definition 3.1.3. The notions of *formula context* and *subformula* are defined in the same way as in the propositional case.

3.2 A Deep Symmetric System

The rules of system SKSgq, a symmetric system for predicate logic, are shown in Figure 3.1. The first and last column show the rules that deal with quantifiers, in the middle there are the rules for the propositional fragment. The u↓ rule corresponds to the R∀ rule in GS1, shown in Figure 3.2. Going up, the R∀ rule removes a universal quantifier from a formula to allow other rules to access this formula. In system SKSgq, inference rules apply deep inside formulas, so there is no need to remove the quantifier. It suffices to move it out of the way using the u↓ rule:

$$\mathsf{u}{\downarrow}\ \frac{\dfrac{S\{\forall x[R,T]\}}{S[\forall xR,\exists xT]}}{S[\forall xR,T]} \quad \text{if } x \text{ is not free in } T,$$

and it vanishes once the proof is complete:

$$= \frac{\mathsf{t}}{\forall x\mathsf{t}} \quad .$$

As a result, the premise of the u↓ rule implies its conclusion, which is not true for the R∀ rule of the sequent calculus. The R∀ rule is the only rule in GS1 with such bad behaviour. In all the rules that I presented in the calculus of structures the premise implies the conclusion.

The instantiation rule n↓ corresponds to R∃. As usual, the substitution operation requires τ to be free for x in R: quantifiers in R do not capture variables in τ. The term τ is not required to be free for x in $S\{R\}$: quantifiers in S may capture variables in τ.

The rules u↑ and n↑ are just the duals of the two rules explained above. They ensure that the system is symmetric.

As in the propositional case we have the deduction theorem:

Theorem 3.2.1 (Deduction Theorem).

There is a derivation $\left.\begin{array}{c}T\\ \big\| \\ R\end{array}\right.$ SKSgq *if and only if there is a proof* $\left.\begin{array}{c} \big\| \\ [\bar{T},R]\end{array}\right.$ SKSgq .

Proof. (same as in the propositional case, Theorem 2.2.4 on page 14)

A proof Π can be obtained from a given derivation Δ as follows:

$$
\begin{array}{c}
T \\
\Delta \parallel \mathsf{SKSgq} \\
R
\end{array}
\qquad \rightsquigarrow \qquad
\begin{array}{c}
\mathsf{i}\!\downarrow \dfrac{\mathsf{t}}{[\bar{T}, T]} \\
[\bar{T}, \Delta] \parallel \mathsf{SKSgq} \\
[\bar{T}, R]
\end{array}
\qquad ,
$$

and a derivation Δ from a given proof Π as follows:

$$
\begin{array}{c}
\Pi \parallel \mathsf{SKSgq} \\
[\bar{T}, R]
\end{array}
\qquad \rightsquigarrow \qquad
\begin{array}{c}
T \\
(T, \Pi) \parallel \mathsf{SKSgq} \\
\mathsf{s}\,\dfrac{(T, [\bar{T}, R])}{[R, (T, \bar{T})]} \\
\mathsf{i}\!\uparrow \overline{\phantom{[R, (T, \bar{T})]}} \\
R
\end{array}
\qquad .
$$

$\qquad\qquad\qquad\qquad\qquad\qquad\qquad\qquad\qquad\qquad\qquad\qquad\qquad\qquad\square$

In the propositional case the analogue of the above theorem holds for the sequent calculus as well and it is proved in the same way: going from left to right by just adding the negated premise throughout the proof tree, and going from right to left by using a cut.

Theorem 3.2.2. *In system* $\mathsf{GS1p} + \mathsf{Cut}$ *there is a derivation* $\begin{array}{c}\vdash T \\ \bigvee \\ \vdash R\end{array}$ *if*

and only if there is a proof $\begin{array}{c}\bigvee \\ \vdash \bar{T}, R\end{array}$.

However, the above does not hold for predicate logic, i.e. system $\mathsf{GS1} + \mathsf{Cut}$. The direction from left to right fails because the premise of the R$\forall$ rule does not imply its conclusion. The proof for the propositional sequent system does not scale to the sequent system for predicate logic: adding formulas to the context of a derivation can violate the proviso of the R$\forall$ rule. In the calculus of structures, on the other hand, the proof for the propositional system scales to the one for predicate logic, as witnessed above. The reason is that the provisos for the equations of variable renaming and vacuous quantifier are local, in the sense that they only require checking the subformula that is being changed, while the proviso of the R$\forall$ rule is global, in the sense that the entire context has to be checked.

$$\text{R}\exists \frac{\vdash \Phi, A[x/\tau]}{\vdash \Phi, \exists x A} \qquad \text{R}\forall \frac{\vdash \Phi, A[x/y]}{\vdash \Phi, \forall x A}$$

Proviso: y is not free in the conclusion of R$\forall$.

Figure 3.2: Quantifier rules of GS1

3.3 Correspondence to the Sequent Calculus

We extend the translations between SKSg and GS1p to translations between SKSgq and GS1. System GS1 is system GS1p extended by the rules shown in Figure 3.2.

The functions $\underline{\quad}_S$ and $\underline{\quad}_G$ are extended in the obvious way:

$$\underline{\exists x A}_S = \exists x \underline{A}_S \qquad \underline{\exists x S}_G = \exists x \underline{S}_G$$
$$\underline{\forall x A}_S = \forall x \underline{A}_S \quad \text{and} \quad \underline{\forall x S}_G = \forall x \underline{S}_G$$

From Sequent Calculus to Calculus of Structures

Theorem 3.3.1.

For every derivation $\overset{\Sigma_1 \ \cdots \ \Sigma_h}{\underset{\Sigma}{\bigtriangledown}}$ *in* GS1 + Cut, *in which the free variables in the premises that are introduced by* R$\forall$ *instances are* $x_1, \ldots, x_n$, *there exists a derivation* $\overset{\forall x_1 \ldots \forall x_n (\underline{\Sigma_1}_S, \ldots, \underline{\Sigma_h}_S)}{\underset{\underline{\Sigma}_S}{\Big\| \text{SKSgq} \setminus \{\text{w}\uparrow, \text{c}\uparrow, \text{u}\uparrow, \text{n}\uparrow\}}}$ *with the same number of cuts.*

Proof. The proof is an extension of the proof of Theorem 2.3.3. There are two more inductive cases, one for R$\exists$, which is easily translated into an n$\downarrow$, and one for R$\forall$, which is shown here:

$$\text{R}\forall \frac{\overset{\Sigma_1 \ \cdots \ \Sigma_{h'}}{\bigtriangledown}}{\vdash \Phi, A[x/y]} \frac{\vdash \Phi, A[x/y]}{\vdash \Phi, \forall x A}$$

By induction hypothesis we have the derivation

$$\forall x_1 \ldots \forall x_{n'}(\underline{\Sigma_1}_s, \ldots, \underline{\Sigma_{h'}}_s)$$

$$\Delta \, \big\| \, \mathsf{SKSgq} \setminus \{\mathsf{w}{\uparrow},\mathsf{c}{\uparrow},\mathsf{u}{\uparrow},\mathsf{n}{\uparrow}\} \qquad ,$$

$$[\underline{\Phi}_s, \underline{A[x/y]}_s]$$

from which we build

$$\forall y \forall x_1 \ldots \forall x_{n'}(\underline{\Sigma_1}_s, \ldots, \underline{\Sigma_{h'}}_s)$$

$$\forall y\{\Delta\} \, \big\| \, \mathsf{SKSgq} \setminus \{\mathsf{w}{\uparrow},\mathsf{c}{\uparrow},\mathsf{u}{\uparrow},\mathsf{n}{\uparrow}\}$$

$$\mathsf{u}{\downarrow} \, \frac{\forall y[\underline{\Phi}_s, \underline{A[x/y]}_s]}{\dfrac{[\exists y \underline{\Phi}_s, \forall y \underline{A[x/y]}_s]}{\dfrac{[\underline{\Phi}_s, \forall y \underline{A[x/y]}_s]}{[\underline{\Phi}_s, \forall x \underline{A}_s]}}} \qquad ,$$

where in the lower instance of the equivalence rule y is not free in $\forall x \underline{A}_s$ and in the upper instance of the equivalence rule y is not free in $\underline{\Phi}_s$: both due to the proviso of the R$\forall$ rule.   □

Since proofs are special derivations, we obtain the following corollaries:

Corollary 3.3.2. *If a sequent Σ has a proof in $\mathsf{GS1} + \mathsf{Cut}$ then $\underline{\Sigma}_s$ has a proof in $\mathsf{SKSgq} \setminus \{\mathsf{w}{\uparrow},\mathsf{c}{\uparrow},\mathsf{u}{\uparrow},\mathsf{n}{\uparrow}\}$.*

Corollary 3.3.3. *If a sequent Σ has a proof in $\mathsf{GS1}$ then $\underline{\Sigma}_s$ has a proof in $\mathsf{SKSgq} \setminus \{\mathsf{i}{\uparrow},\mathsf{w}{\uparrow},\mathsf{c}{\uparrow},\mathsf{u}{\uparrow},\mathsf{n}{\uparrow}\}$.*

From Calculus of Structures to Sequent Calculus

As in the propositional case, we need the following lemma to imitate deep inference in the sequent calculus:

Lemma 3.3.4. *For every two formulas A, B and every formula context $C\{\ \}$*

there exists a derivation
$$\begin{array}{c} \vdash A, \bar{B} \\ \diagdown\diagup \\ \vdash C\{A\}, \overline{C\{B\}} \end{array}$$
 in $\mathsf{GS1}$.

Proof. There are two cases needed in addition to the proof of Lemma 2.3.7: $C\{\ \} = \exists x C'\{\ \}$ and $C\{\ \} = \forall x C'\{\ \}$. The first case is shown here, the second is similar:

$$\dfrac{\vdash A, \bar{B}}{\nabla_{\!\!\Delta}}$$

$$\mathsf{R\exists}\ \dfrac{\vdash C'\{A\}, \overline{C'\{B\}}}{\mathsf{R\forall}\ \dfrac{\vdash \exists x C'\{A\}, \overline{C'\{B\}}}{\vdash \exists x C'\{A\}, \forall x \overline{C'\{B\}}}}\quad,$$

where Δ exists by induction hypothesis. □

Now we can translate derivations in system SKSgq into derivations in the sequent calculus:

Theorem 3.3.5. *For every derivation* $\begin{matrix} Q \\ \Big\| \text{SKSgq} \\ P \end{matrix}$ *there exists a derivation* $\dfrac{\vdash \underline{Q}_G}{\nabla}{\vdash \underline{P}_G}$

in GS1 + Cut.

Proof. The proof is an extension of the proof of Theorem 2.3.8 on page 18. The base cases are the same, in the inductive cases the existence of Δ_1 follows from Lemma 3.3.4. Corresponding to the rules for quantifiers, there are four additional inductive cases. For the rules

$$\mathsf{u}{\downarrow}\ \dfrac{S\{\forall x[R,T]\}}{S[\forall xR, \exists xT]} \qquad \text{and} \qquad \mathsf{u}{\uparrow}\ \dfrac{S(\exists xR, \forall xT)}{S\{\exists x(R,T)\}}$$

we have the proofs

$$\mathsf{R\vee}\ \dfrac{\mathsf{R\forall}\ \dfrac{\mathsf{R\exists}\ \dfrac{\mathsf{R\exists}\ \dfrac{\mathsf{R\wedge}\ \dfrac{\mathsf{Ax}\ \dfrac{}{\vdash R, \bar{R}}\quad \mathsf{Ax}\ \dfrac{}{\vdash T, \bar{T}}}{\vdash R, T, \bar{R} \wedge \bar{T}}}{\vdash R, \exists xT, \bar{R} \wedge \bar{T}}}{\vdash R, \exists xT, \exists x(\bar{R} \wedge \bar{T})}}{\vdash \forall xR, \exists xT, \exists x(\bar{R} \wedge \bar{T})}}{\vdash \forall xR \vee \exists xT, \exists x(\bar{R} \wedge \bar{T})} \qquad \text{and}$$

$$\mathsf{R\vee}\ \dfrac{\mathsf{R\forall}\ \dfrac{\mathsf{R\exists}\ \dfrac{\mathsf{R\exists}\ \dfrac{\mathsf{R\wedge}\ \dfrac{\mathsf{Ax}\ \dfrac{}{\vdash R, \bar{R}}\quad \mathsf{Ax}\ \dfrac{}{\vdash T, \bar{T}}}{\vdash R \wedge T, \bar{R}, \bar{T}}}{\vdash R \wedge T, \bar{R}, \exists x\bar{T}}}{\vdash \exists x(R \wedge T), \bar{R}, \exists x\bar{T}}}{\vdash \exists x(R \wedge T), \forall x\bar{R}, \exists x\bar{T}}}{\vdash \exists x(R \wedge T), \forall x\bar{R} \vee \exists x\bar{T}}\quad,$$

and for the rules

$$\mathsf{n}{\downarrow}\ \dfrac{S\{R[x/\tau]\}}{S\{\exists xR\}} \qquad \text{and} \qquad \mathsf{n}{\uparrow}\ \dfrac{S\{\forall xR\}}{S\{R[x/\tau]\}}$$

we have the proofs

$$\mathsf{R\exists}\ \dfrac{\mathsf{Ax}\ \dfrac{}{\vdash R[x/\tau], \overline{R[x/\tau]}}}{\vdash \exists xR, \overline{R[x/\tau]}} \qquad \text{and} \qquad \mathsf{R\exists}\ \dfrac{\mathsf{Ax}\ \dfrac{}{\vdash R[x/\tau], \bar{R}[x/\tau]}}{\vdash R[x/\tau], \exists x\bar{R}}\quad.$$

$$\mathsf{i}{\downarrow}\ \frac{S\{\mathsf{t}\}}{S[R,\bar{R}]} \qquad \mathsf{w}{\downarrow}\ \frac{S\{\mathsf{f}\}}{S\{R\}} \qquad \mathsf{c}{\downarrow}\ \frac{S[R,R]}{S\{R\}}$$

$$\mathsf{s}\ \frac{S([R,T],U)}{S[(R,U),T]} \qquad \mathsf{u}{\downarrow}\ \frac{S\{\forall x[R,T]\}}{S[\forall xR,\exists xT]} \qquad \mathsf{n}{\downarrow}\ \frac{S\{R[x/\tau]\}}{S\{\exists xR\}}$$

Figure 3.3: System KSgq

$\square$

Corollary 3.3.6. *If a formula S has a proof in* SKSgq *then* $\vdash \underline{S}_{G}$ *has a proof in* GS1.

Soundness and completeness of SKSgq, i.e. the fact that a formula has a proof in SKSgq if and only if it is valid, follows from soundness and completeness of GS1 by Corollaries 3.3.2 and 3.3.6. Moreover, a formula T implies a formula R if and only if there is a derivation from T to R, which follows from soundness and completeness and the deduction theorem.

3.4   Cut Admissibility

Just like in the propositional case, the up-rules of the symmetric system are admissible. By removing them from SKSgq we obtain the asymmetric, cut-free system shown in Figure 3.3, which is called system KSgq.

Theorem 3.4.1. *The rules* $\mathsf{i}{\uparrow}$, $\mathsf{w}{\uparrow}$, $\mathsf{c}{\uparrow}$, $\mathsf{u}{\uparrow}$ *and* $\mathsf{n}{\uparrow}$ *are admissible for system* KSgq.

Proof.

$\square$

Corollary 3.4.2. *The systems* SKSgq *and* KSgq *are equivalent.*

Chapter Summary

We have extended the deep symmetric system of the previous chapter to predicate logic. The translations to and from the the one-sided sequent calculus have also been extended—and thus the proof of cut admissibility. A nice feature of this system is that, for all its rules, the premise implies the conclusion, which is not true in sequent systems for predicate logic.

Chapter 4

Locality

Inference rules that copy an unbounded quantity of information are problematic from the points of view of complexity and implementation. In the sequent calculus, an example is given by the contraction rule:

$$\frac{\vdash \Phi, A, A}{\vdash \Phi, A} \ .$$

Here, going from bottom to top in constructing a proof, a formula A of unbounded size is duplicated. Whatever mechanism performs this duplication, it has to inspect all of A, so it has to have a *global* view on A. If, for example, we had to implement contraction on a distributed system, where each processor has a limited amount of local memory, the formula A could be spread over a number of processors. In that case, no single processor has a global view on A, and we should put in place complex mechanisms to cope with the situation.

Let us call *local* those inference rules that do not require such a global view on formulas of unbounded size, and *non-local* those rules that do. Further examples of non-local rules are the promotion rule in the sequent calculus for linear logic (left, [18]) and context-sharing (or additive) rules found in various sequent systems (right, [44]):

$$\frac{\vdash A, ?B_1, \ldots, ?B_n}{\vdash !A, ?B_1, \ldots, ?B_n} \quad \text{and} \quad \frac{\vdash \Phi, A \quad \vdash \Phi, B}{\vdash \Phi, A \wedge B} \ .$$

To apply the promotion rule, one has to check whether all formulas in the context are prefixed with a question mark modality: the number of formulas to check is unbounded. To apply the context-sharing R∧ rule, a context of unbounded size has to be copied.

While there are methods to solve these problems in an implementation, an interesting question is whether it is possible to approach them proof-theoretically, i.e. by avoiding non-local rules. This chapter gives an affirmative answer by presenting systems for both classical propositional and first-order predicate logic in which context-sharing rules as well as contraction are replaced by local rules. For propositional logic it is even possible to obtain a system which contains local rules only.

Locality is achieved by reducing the problematic rules to their atomic forms. This phenomenon is not restricted to the calculus of structures: in most sequent systems for classical logic the identity axiom is reduced to its atomic form, i.e.

$$\overline{\vdash A, \bar{A}} \qquad \text{is equivalently replaced by} \qquad \overline{\vdash a, \bar{a}} \quad,$$

where a is an atom. Contraction, however, cannot be replaced by its atomic form in known sequent systems, as we will see in Section 4.3.

In this chapter we will see a local system for propositional logic and a system for predicate logic which is local except for the treatment of variables.

4.1 Propositional Logic

In the following we will obtain system SKS, which is equivalent to system SKSg, but identity, cut, weakening and contraction are restricted to atomic form. This entails locality of the system.

4.1.1 Reducing Rules to Atomic Form

In the sequent calculus, the identity rule can be reduced to its atomic form. The same is true for our system, i.e.

$$\mathsf{i}{\downarrow}\,\frac{S\{\mathsf{t}\}}{S[R,\bar{R}]} \qquad \text{is equivalently replaced by} \qquad \mathsf{ai}{\downarrow}\,\frac{S\{\mathsf{t}\}}{S[a,\bar{a}]} \quad,$$

where $\mathsf{ai}{\downarrow}$ is the *atomic identity* rule. Similarly to the sequent calculus, this is achieved by inductively replacing an instance of the general identity rule by instances on smaller formulas:

$$\mathsf{i}{\downarrow}\,\frac{S\{\mathsf{t}\}}{S[P,Q,(\bar{P},\bar{Q})]} \quad \leadsto \quad \mathsf{s}\,\frac{\mathsf{s}\,\dfrac{\mathsf{i}{\downarrow}\,\dfrac{S\{\mathsf{t}\}}{S[Q,\bar{Q}]}}{S([P,\bar{P}],[Q,\bar{Q}])}}{\dfrac{S[Q,([P,\bar{P}],\bar{Q})]}{S[P,Q,(\bar{P},\bar{Q})]}} \quad.$$

What is new in the calculus of structures is that the cut can also be reduced to atomic form: just take the dual derivation of the one above:

$$
\mathsf{i}{\uparrow}\,\frac{S(\bar{P},\bar{Q},[P,Q])}{S\{\mathsf{f}\}}
\qquad \rightsquigarrow \qquad
\mathsf{i}{\uparrow}\,\frac{\displaystyle \mathsf{s}\,\frac{\displaystyle \mathsf{s}\,\frac{\displaystyle S(\bar{Q},\bar{P},[Q,P])}{S(\bar{Q},[(\bar{P},P),Q])}}{S[(\bar{P},P),(\bar{Q},Q)]}}{\displaystyle \mathsf{i}{\uparrow}\,\frac{S(\bar{Q},Q)}{S\{\mathsf{f}\}}}
\quad .
$$

In this way, the general cut rule is equivalently replaced by the *atomic cut* rule:

$$
\mathsf{i}{\uparrow}\,\frac{S(R,\bar{R})}{S\{\mathsf{f}\}}
\qquad \text{is equivalently replaced by} \qquad
\mathsf{ai}{\uparrow}\,\frac{S(a,\bar{a})}{S\{\mathsf{f}\}}
\quad .
$$

It turns out that weakening can also be reduced to atomic form. When identity, cut and weakening are restricted to atomic form, there is only one non-local rule left in system KSg: contraction. It can not be reduced to atomic form in system KSg. Tiu solved this problem when he discovered the *medial* rule [8]:

$$
\mathsf{m}\,\frac{S[(R,U),(T,V)]}{S([R,T],[U,V])}
\quad .
$$

This rule has no analogue in the sequent calculus. But it is clearly sound because we can derive it:

Proposition 4.1.1. *The medial rule is derivable for* $\{\mathsf{c}{\downarrow},\mathsf{w}{\downarrow}\}$. *Dually, the medial rule is derivable for* $\{\mathsf{c}{\uparrow},\mathsf{w}{\uparrow}\}$.

Proof. The medial rule is derivable as follows (or dually):

$$
\mathsf{c}{\downarrow}\,\frac{\displaystyle \mathsf{w}{\downarrow}\,\frac{\displaystyle \mathsf{w}{\downarrow}\,\frac{\displaystyle \mathsf{w}{\downarrow}\,\frac{\displaystyle \mathsf{w}{\downarrow}\,\frac{S[(R,U),(T,V)]}{S[(R,U),(T,[U,V])]}}{S[(R,U),([R,T],[U,V])]}}{S[(R,[U,V]),([R,T],[U,V])]}}{S[([R,T],[U,V]),([R,T],[U,V])]}}{S([R,T],[U,V])}
\quad .
$$

$\square$

The medial rule has also been considered by Došen and Petrić as a composite arrow in the free bicartesian category, cf. the end of Section 4 in [13]. It is composed of four projections and a pairing of identities (or dually) in the same way as medial is derived using four weakenings and a contraction in the proof above.

Once we admit medial, then not only identity, cut and weakening, but also contraction is reducible to atomic form:

Theorem 4.1.2. *The rules* $i{\downarrow}$, $w{\downarrow}$ *and* $c{\downarrow}$ *are derivable for* $\{ai{\downarrow},s\}$, $\{aw{\downarrow},s\}$ *and* $\{ac{\downarrow},m\}$, *respectively. Dually, the rules* $i{\uparrow}$, $w{\uparrow}$ *and* $c{\uparrow}$ *are derivable for* $\{ai{\uparrow},s\}$, $\{aw{\uparrow},s\}$ *and* $\{ac{\uparrow},m\}$, *respectively.*

Proof. I will show derivability of the rules $\{i{\downarrow},w{\downarrow},c{\downarrow}\}$ for the respective systems. The proof of derivability of their co-rules is dual.

Given an instance of one of the following rules:

$$i{\downarrow}\,\frac{S\{t\}}{S[R,\bar{R}]} \quad , \qquad w{\downarrow}\,\frac{S\{f\}}{S\{R\}} \quad , \qquad c{\downarrow}\,\frac{S[R,R]}{S\{R\}} \quad ,$$

construct a new derivation by structural induction on R:

1. R is an atom. Then the instance of the general rule is also an instance of its atomic form.

2. $R = t$ or $R = f$. Then the instance of the general rule is an instance of the equivalence rule, with the only exception of weakening in case that $R = t$. Then this instance of weakening can be replaced by

$$= \frac{S\{f\}}{\dfrac{S([t,t],f)}{\quad}}$$
$$s\,\frac{S([t,t],f)}{\dfrac{S[t,(t,f)]}{S\{t\}}} \quad .$$

3. $R = [P,Q]$. Apply the induction hypothesis respectively on

$$i{\downarrow}\,\frac{\dfrac{\quad}{i{\downarrow}\,\dfrac{S\{t\}}{S[Q,\bar{Q}]}}}{\dfrac{S([P,\bar{P}],[Q,\bar{Q}])}{\dfrac{s\,\dfrac{S[Q,([P,\bar{P}],\bar{Q})]}{S[P,Q,(\bar{P},\bar{Q})]}}{\quad}}} \quad , \qquad w{\downarrow}\,\frac{\dfrac{= \dfrac{S\{f\}}{S[f,f]}}{w{\downarrow}\,\dfrac{S[f,Q]}{S[P,Q]}}}{\quad} \quad , \qquad c{\downarrow}\,\frac{c{\downarrow}\,\dfrac{S[P,P,Q,Q]}{S[P,P,Q]}}{S[P,Q]} \quad .$$

$$\text{ai}\downarrow \frac{S\{\mathsf{t}\}}{S[a,\bar{a}]} \qquad\qquad \text{ai}\uparrow \frac{S(a,\bar{a})}{S\{\mathsf{f}\}}$$

$$\mathsf{s}\, \frac{S([R,U],T)}{S[(R,T),U]}$$

$$\mathsf{m}\, \frac{S[(R,U),(T,V)]}{S([R,T],[U,V])}$$

$$\text{aw}\downarrow \frac{S\{\mathsf{f}\}}{S\{a\}} \qquad\qquad \text{aw}\uparrow \frac{S\{a\}}{S\{\mathsf{t}\}}$$

$$\text{ac}\downarrow \frac{S[a,a]}{S\{a\}} \qquad\qquad \text{ac}\uparrow \frac{S\{a\}}{S(a,a)}$$

Figure 4.1: System SKS

4. $R = (P,Q)$. Apply the induction hypothesis respectively on

$$\mathsf{s}\,\frac{\mathsf{s}\,\dfrac{\mathsf{i}\downarrow\dfrac{\mathsf{i}\downarrow\dfrac{S\{\mathsf{t}\}}{S[Q,\bar{Q}]}}{S([P,\bar{P}],[Q,\bar{Q}])}}{S([(P,\bar{P}],Q),\bar{Q}]}}{S[(P,Q),\bar{P},\bar{Q}]}\quad,\qquad \mathsf{w}\downarrow\frac{\mathsf{w}\downarrow\dfrac{=\dfrac{S\{\mathsf{f}\}}{S(\mathsf{f},\mathsf{f})}}{S(\mathsf{f},Q)}}{S(P,Q)}\quad,\qquad \mathsf{c}\downarrow\frac{\mathsf{c}\downarrow\dfrac{\mathsf{m}\dfrac{S[(P,Q),(P,Q)]}{S([P,P],[Q,Q])}}{S([P,P],Q)}}{S(P,Q)}\quad.$$

$\square$

4.1.2   A Local System for Propositional Logic

We now obtain the local system SKS from SKSg by restricting identity, cut, weakening and contraction to atomic form and adding medial. It is shown in Figure 4.1. The names of the rules are as in system SKSg, except that the atomic rules carry the attribute *atomic*, for example aw↑ is the *atomic co-weakening* rule:

Theorem 4.1.3. *System* SKS *and system* SKSg *are strongly equivalent.*

Proof. Derivations in SKSg are translated to derivations in SKS by Theorem 4.1.2, and vice versa by Proposition 4.1.1.   □

Thus, all results obtained for the non-local system, in particular the correspondence with the sequent calculus and admissibility of the up-rules, also hold for the local system. By removing the up-rules from system SKS we obtain system KS, shown in Figure 4.2.

$$\mathsf{ai}{\downarrow}\,\frac{S\{\mathsf{t}\}}{S[a,\bar{a}]} \qquad \mathsf{aw}{\downarrow}\,\frac{S\{\mathsf{f}\}}{S\{a\}} \qquad \mathsf{ac}{\downarrow}\,\frac{S[a,a]}{S\{a\}}$$

$$\mathsf{s}\,\frac{S([R,T],U)}{S[(R,U),T]} \qquad \mathsf{m}\,\frac{S[(R,T),(U,V)]}{S([R,U],[T,V])}$$

Figure 4.2: System KS

Theorem 4.1.4. *System* KS *and system* KSg *are strongly equivalent.*

Proof. As in the proof of the previous theorem (Theorem 4.1.3).   □

Notation 4.1.5. While the non-local rules, general identity, weakening, contraction and their duals $\{\mathsf{i}{\downarrow},\mathsf{i}{\uparrow},\mathsf{w}{\downarrow},\mathsf{w}{\uparrow},\mathsf{c}{\downarrow},\mathsf{c}{\uparrow}\}$ do not belong to SKS, I will freely use them to denote a corresponding derivation in SKS according to Theorem 4.1.2. For example, I will use

$$\mathsf{c}{\downarrow}\,\frac{[(a,b),(a,b)]}{(a,b)}$$

to denote either

$$\mathsf{ac}{\downarrow}\,\frac{\mathsf{m}\,\dfrac{[(a,b),(a,b)]}{\dfrac{([a,a],[b,b])}{\dfrac{([a,a],b)}{(a,b)}}}}{} \qquad \text{or} \qquad \mathsf{ac}{\downarrow}\,\frac{\mathsf{m}\,\dfrac{[(a,b),(a,b)]}{\dfrac{([a,a],[b,b])}{\dfrac{(a,[b,b])}{(a,b)}}}}{} \quad .$$

In system SKSg and also in sequent systems, there is no bound on the size of formulas that can appear as an active formula in an instance of the contraction rule. Implementing those systems for proof search thus requires duplicating formulas of unbounded size. One could avoid this by putting in place some mechanism of sharing and copying on demand, but this would make for a significant difference between the formal system and its implementation. It is more desirable to have no such difference between the formal system that is studied theoretically and its implementation that is used.

In system SKS, no rule requires duplicating formulas of unbounded size. In fact, because no rule needs to inspect formulas of unbounded size, I call this system *local*. The atomic rules only need to duplicate, erase or compare atoms. The switch rule involves formulas of unbounded size, namely R, T and U. But it does not require inspecting them. To see this, consider formulas represented as binary trees in the obvious way. Then the switch rule can be implemented by changing the marking of two nodes and exchanging two pointers:

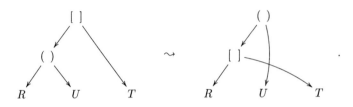

The same technique works for medial. The equations are local as well, including the De Morgan laws. However, since the rules in SKS introduce negation only on atoms, it is even possible to restrict negation to atoms from the beginning, as is customary in the one-sided sequent calculus, and drop the equations for negation entirely.

The concept of locality depends on the representation of formulas. Rules that are local for one representation may not be local when another representation is used. For example, the switch rule is local when formulas are represented as trees, but it is not local when formulas are represented as strings.

One motivation for locality is to simplify distributed implementation of an inference system. Of course, locality by itself still makes no distributed implementation. There are tasks to accomplish in an implementation of an inference system that in general require a global view on formulas, for example matching a rule, i.e. finding a redex. There should also be some mechanism for backtracking. I do not see how these problems can be approached within a proof-theoretic system with properties like cut elimination. However, the application of a rule, i.e. producing the contractum from the redex, is achieved locally in system SKS. For that reason I believe that it lends itself more easily to distributed implementation than other systems.

4.2   Predicate Logic

For predicate logic we will now obtain system SKSq, which is equivalent to
system SKSgq, but, like in the propositional case, cut, identity, weakening
and contraction are restricted to atomic form. The resulting system is local
except for the rules that instantiate variables or check for free occurrences
of a variable.

4.2.1   Reducing Rules to Atomic Form

Cut and identity are reduced to atomic form by using the rules $u\!\downarrow$ and $u\!\uparrow$,
which follow a scheme or recipe due to Guglielmi [21]. This scheme, which
also yields the switch rule, ensures atomicity of cut and identity not only
for classical logic but also for several other logics.

To reduce contraction to atomic form, we need the following rules in addition
to medial:

$$l_1\!\downarrow \frac{S[\exists x R, \exists x T]}{S\{\exists x[R,T]\}} \qquad l_1\!\uparrow \frac{S\{\forall x(R,T)\}}{S(\forall x R, \forall x T)}$$

$$l_2\!\downarrow \frac{S[\forall x R, \forall x T]}{S\{\forall x[R,T]\}} \qquad l_2\!\uparrow \frac{S\{\exists x(R,T)\}}{S(\exists x R, \exists x T)} \qquad .$$

Note that they are local. Like medial, they have no analogues in the sequent
calculus. In system SKSgq, and similarly in the sequent calculus, the corresponding inferences are made using contraction and weakening:

Proposition 4.2.1. *The rules* $\{l_1\!\downarrow, l_2\!\downarrow\}$ *are derivable for* $\{c\!\downarrow, w\!\downarrow\}$. *Dually,*
the rules $\{l_1\!\uparrow, l_2\!\uparrow\}$ *are derivable for* $\{c\!\uparrow, w\!\uparrow\}$.

Proof. I show the case for $l_1\!\downarrow$, the other cases are similar or dual:

$$\begin{array}{c} w\!\downarrow \dfrac{S[\exists x R, \exists x T]}{S[\exists x R, \exists x[R,T]]} \\ w\!\downarrow \dfrac{}{S[\exists x[R,T], \exists x[R,T]]} \\ c\!\downarrow \dfrac{}{S\{\exists x[R,T]\}} \end{array} \qquad .$$

$\square$

Theorem 4.2.2. *The rules* $i\!\downarrow$, $w\!\downarrow$ *and* $c\!\downarrow$ *are derivable for* $\{ai\!\downarrow, s, u\!\downarrow\}$, $\{aw\!\downarrow, s\}$
and $\{ac\!\downarrow, m, l_1\!\downarrow, l_2\!\downarrow\}$, *respectively. Dually, the rules* $i\!\uparrow$, $w\!\uparrow$ *and* $c\!\uparrow$ *are deriv-*
able for $\{ai\!\uparrow, s, u\!\uparrow\}$, $\{aw\!\uparrow, s\}$ *and* $\{ac\!\uparrow, m, l_1\!\uparrow, l_2\!\uparrow\}$, *respectively.*

Proof. The proof is an extension of the proof of Theorem 4.1.2 by the inductive cases for the quantifiers. Given an instance of one of the following rules:

$$\text{i}\!\downarrow \frac{S\{\text{t}\}}{S[R,\bar{R}]} \quad , \quad \text{w}\!\downarrow \frac{S\{\text{f}\}}{S\{R\}} \quad , \quad \text{c}\!\downarrow \frac{S[R,R]}{S\{R\}} \quad ,$$

construct a new derivation by structural induction on R:

1. $R = \exists x T$. Apply the induction hypothesis respectively on

$$\text{i}\!\downarrow \frac{=\dfrac{S\{\text{t}\}}{S\{\forall x \text{t}\}}}{\text{u}\!\downarrow \dfrac{S\{\forall x[T,\bar{T}]\}}{S[\exists x T, \forall x \bar{T}]}} \quad , \quad \text{w}\!\downarrow \frac{=\dfrac{S\{\text{f}\}}{S\{\exists x \text{f}\}}}{S\{\exists x T\}} \quad , \quad \text{c}\!\downarrow \frac{\text{l}_1\!\downarrow \dfrac{S[\exists x T, \exists x T]}{S\{\exists x[T,T]\}}}{S\{\exists x T\}} \quad .$$

2. $R = \forall x T$. Apply the induction hypothesis respectively on

$$\text{i}\!\downarrow \frac{=\dfrac{S\{\text{t}\}}{S\{\forall x \text{t}\}}}{\text{u}\!\downarrow \dfrac{S\{\forall x[T,\bar{T}]\}}{S[\forall x T, \exists x \bar{T}]}} \quad , \quad \text{w}\!\downarrow \frac{=\dfrac{S\{\text{f}\}}{S\{\forall x \text{f}\}}}{S\{\forall x T\}} \quad , \quad \text{c}\!\downarrow \frac{\text{l}_2\!\downarrow \dfrac{S[\forall x T, \forall x T]}{S\{\forall x[T,T]\}}}{S\{\forall x T\}} \quad .$$

$\square$

4.2.2 A Local System for Predicate Logic

We now obtain system SKSq from SKSgq by restricting identity, cut, weakening and contraction to atomic form and adding the rules $\{\text{m},\text{l}_1\!\downarrow,\text{l}_2\!\downarrow,\text{l}_1\!\uparrow,\text{l}_2\!\uparrow\}$. It is shown in Figure 4.3.

As in all the systems considered, the up-rules, i.e. $\{\text{n}\!\uparrow,\text{u}\!\uparrow,\text{l}_1\!\uparrow,\text{l}_2\!\uparrow\}$ are admissible. Hence, system KSq, shown in Figure 4.4, is complete.

Theorem 4.2.3. *System* SKSq *and system* SKSgq *are strongly equivalent. Also, system* KSq *and system* KSgq *are strongly equivalent.*

Proof. Derivations in SKSgq are translated to derivations in SKSq by Theorem 4.2.2, and vice versa by Proposition 4.2.1. The same holds for KSgq and KSq. $\square$

Thus, all results obtained for system SKSgq also hold for system SKSq. As in the propositional case, I will freely use general identity, cut, weakening

$$ai{\downarrow} \frac{S\{t\}}{S[a,\bar{a}]} \qquad\qquad ai{\uparrow} \frac{S(a,\bar{a})}{S\{f\}}$$

$$s\,\frac{S([R,U],T)}{S[(R,T),U]}$$

$$u{\downarrow} \frac{S\{\forall x[R,T]\}}{S[\forall xR,\exists xT]} \qquad\qquad u{\uparrow} \frac{S(\exists xR,\forall xT)}{S\{\exists x(R,T)\}}$$

$$l_1{\downarrow} \frac{S[\exists xR,\exists xT]}{S\{\exists x[R,T]\}} \qquad\qquad l_1{\uparrow} \frac{S\{\forall x(R,T)\}}{S(\forall xR,\forall xT)}$$

$$m\,\frac{S[(R,U),(T,V)]}{S([R,T],[U,V])}$$

$$l_2{\downarrow} \frac{S[\forall xR,\forall xT]}{S\{\forall x[R,T]\}} \qquad\qquad l_2{\uparrow} \frac{S\{\exists x(R,T)\}}{S(\exists xR,\exists xT)}$$

$$aw{\downarrow} \frac{S\{f\}}{S\{a\}} \qquad\qquad aw{\uparrow} \frac{S\{a\}}{S\{t\}}$$

$$ac{\downarrow} \frac{S[a,a]}{S\{a\}} \qquad\qquad ac{\uparrow} \frac{S\{a\}}{S(a,a)}$$

$$n{\downarrow} \frac{S\{R[x/\tau]\}}{S\{\exists xR\}} \qquad\qquad n{\uparrow} \frac{S\{\forall xR\}}{S\{R[x/\tau]\}}$$

Figure 4.3: System SKSq

and contraction to denote a corresponding derivation in SKSq according to Theorem 4.2.2.

As we have seen in the previous section, the technique of reducing contraction to atomic form to obtain locality also works in the case of predicate logic: the non-local rule c↓ is equivalently replaced by local ones, namely $\{ac{\downarrow}, m, l_1{\downarrow}, l_2{\downarrow}\}$.

However, there are other sources of non-locality in system SKSq. One is the

$$\text{ai}\downarrow \frac{S\{t\}}{S[a,\bar{a}]} \qquad \text{aw}\downarrow \frac{S\{f\}}{S\{a\}} \qquad \text{ac}\downarrow \frac{S[a,a]}{S\{a\}}$$

$$\text{s}\,\frac{S([R,T],U)}{S[(R,U),T]} \qquad \text{u}\downarrow \frac{S\{\forall x[R,T]\}}{S[\forall xR,\exists xT]} \qquad \text{m}\,\frac{S[(R,T),(U,V)]}{S([R,U],[T,V])}$$

$$\text{n}\downarrow \frac{S\{R[x/\tau]\}}{S\{\exists xR\}} \qquad \text{l}_1\downarrow \frac{S[\exists xR,\exists xT]}{S\{\exists x[R,T]\}} \qquad \text{l}_2\downarrow \frac{S[\forall xR,\forall xT]}{S\{\forall x[R,T]\}}$$

Figure 4.4: System KSq

condition on the quantifier equations:

$$\forall yR = \exists yR = R \qquad \text{where } y \text{ is not free in } R.$$

To add or remove a quantifier, a formula of unbounded size has to be checked for occurrences of the variable y.

Another is the n$\downarrow$ rule, in which a term τ of unbounded size has to be copied into an unbounded number of occurrences of x in R. It is non-local for two distinct reasons: 1) the unbounded size of τ and 2) the unbounded number of occurrences of x in R. The unboundedness of term τ can be dealt with, as we will see in Section 5.2.1, page 52. But this does not address the problem of the unbounded number of occurrences of x in R.

Is it possible to obtain a system for first-order predicate logic in which variables are treated locally? I do not know how to do it without adding new symbols to the language of predicate logic. But it is conceivable to obtain a local system by introducing substitution operators together with rules that explicitly handle the instantiation of variables piece by piece. The question is whether this can be done without losing the good properties, especially cut elimination and simplicity.

4.3 Locality and the Sequent Calculus

The question is whether deep inference is needed for locality, or whether it can be obtained in sequent systems as well. In system G3cp [44], for example, contraction is admissible and can thus trivially be restricted to atoms or to the bottom of a proof. However, G3cp has an additive (or context-sharing) R∧-rule, so these restrictions on contraction do not entail locality. Contraction is admissible, but additive rules such as R∧ implicitly duplicate formulas which may be non-atomic. Of course, R∧ is not admissible.

It thus remains to be seen whether locality can be achieved in systems with a multiplicative (or context-independent) R∧-rule. It turns out not to be the case because contraction cannot be reduced to atomic form. The argument given below is for system GS1p with multiplicative context treatment, shown in Figure 2.4 on page 15. However, it works for various sequent systems, as long as they have a multiplicative R∧-rule.

Proposition 4.3.1. *There is a valid sequent that has no proof in multiplicative* GS1p *in which all contractions are atomic.*

Proof. Consider the following sequent:

$$\vdash a \wedge b, (\bar{a} \vee \bar{b}) \wedge (\bar{a} \vee \bar{b}) \quad . \tag{4.1}$$

The sequent contains no atoms, so atomic contraction cannot be applied. Each applicable rule leads to a premise that is not valid.   □

Chapter Summary

Starting from the deductive systems that were introduced in the previous two chapters, we have obtained strongly equivalent systems in which identity, cut, weakening and contraction are reduced to their atomic forms. The resulting rules are local, which could be interesting for example for distributed implementation. For propositional logic we obtained a system which is local, for predicate logic one which is local except for the treatment of variables. In the sequent calculus locality can not be achieved because contraction is not reducible to atomic form.

Chapter 5

Finite Choice

I will call *finitely generating* an inference rule if, given its conclusion, there is only a finite number of premises to choose from. Otherwise a rule is called *infinitely generating*. The cut rule in a sequent system is infinitely generating: given its conclusion, there is an infinite choice of premises, corresponding to an infinite choice of cut formulas. Much effort has been devoted to eliminating this source on infinity in various systems: theorems of cut elimination remove infinite choice together with the cut rule itself, and are at the core of proof theory. There is another source of infinite choice in the bottom-up construction of a first order proof, namely the choice in instantiating an existentially quantified variable. Research grounded in Herbrand's theorem [26] deals with this aspect and is at the core of automated deduction and logic programming.

This chapter shows how one can eliminate all sources of infinite choice in a system of first order classical logic in a very simple way. The main idea we exploit is that there are actually two sources of infinite choice in the cut rule: an infinite choice of atoms and an infinite choice in how these atoms can be combined for making formulas. In predicate logic the first source of infinite choice can again be divided into an infinite choice of predicate symbols on one hand and an infinite choice of substitutions on the other hand. Deep inference allows us to separate these various kinds of infinite choice.

In the sequent calculus, it is impossible to separate these kinds of infinite choice in a cut rule without going through cut elimination. Instead, in the calculus of structures one can straightforwardly reduce the cut rule to its atomic form, as we have seen in the previous chapter. This has the advantage of not presenting infinite choice in combining atoms to shape a formula. Similar techniques reduce the instantiation rules into more elementary ones. Infinite choice in the elementary rules so produced can be removed by simple considerations that essentially limit the range of possibilities to the atoms and terms that already appear in the conclusions of rules.

As shown before, systems in the calculus of structures offer the same proof theoretical properties as systems in the sequent calculus, in particular it is possible to prove cut elimination. The point here is that it is possible to eliminate infinite choice without having to use these more complex methods. As an example, we will see how to prove consistency in this setting.

The point here is not that we can obtain finitely generating deductive systems for propositional and predicate logic. There already are plenty of deductive systems like cut-free sequent systems, or methods from automated theorem proving like resolution [33], tableaux [38] and the connection method [45]. These systems are finitely generating (or at least have implementations based on finitely generating rules). This should come as no surprise since these systems and methods are cut-free. Here, we keep the cut and obtain finite choice nevertheless.

The notion of a finitely generating inference rule is closely related to that of an *analytic* rule, cf. Smullyan [37]. An analytic rule is one that obeys the subformula property. We tend to think of the notion of being finitely generating as a more general, weaker subformula property: there are interesting rules that are finitely generating but do not obey the subformula property, for example in system GS4ip, cf. Dyckhoff [14]. However, not all analytic rules are finitely generating, as witnessed by the existential-right rule. This is due to the fact that analyticity is defined with respect to the notion of Gentzen subformula (where instances of subformulae count as subformulae), rather than the literal notion of subformula.

In the systems presented here, we have an analytic cut rule: it only introduces atoms that occur in the conclusion. So, even when allowing the use of cut, the only infinity that remains in proof search is the unboundedness of the proofs themselves: at any given step, there are finitely many inferences possible, and each inference rule can only be applied in a finite number of different ways. The cut behaves similarly to the contraction rule: it is always applicable, but only in a finite number of ways.

5.1   Propositional Logic

5.1.1   Eliminating Infinite Choice in Inference Rules

There are two infinitely generating rules in system SKS: the atomic co-weakening and the atomic cut rule. In the following we will see how to replace those rules by finitely generating ones without affecting provability. The equivalence rule is also infinitely generating, but can be broken up into several rules, for commutativity, associativity, and so on. Those rules are finitely generating.

The Atomic Co-weakening Rule

The rule aw↑ is clearly infinitely generating since there is an infinite choice of atoms, but it can immediately be eliminated by using a cut and an instance of aw↓ as follows:

$$
\mathsf{aw}{\uparrow}\,\frac{S\{a\}}{S\{\mathsf{t}\}} \qquad \rightsquigarrow \qquad \mathsf{aw}{\downarrow}\cfrac{\mathsf{s}\cfrac{=\cfrac{S\{a\}}{S(a,[\mathsf{t},\mathsf{f}])}}{S[\mathsf{t},(a,\mathsf{f})]}}{\mathsf{ai}{\uparrow}\cfrac{S[\mathsf{t},(a,\bar{a})]}{=\cfrac{S[\mathsf{t},\mathsf{f}]}{S\{\mathsf{t}\}}}} \qquad .
$$

The Atomic Cut Rule

The cut is the most prominent infinitely generating rule. The first source of infinite choice, the arbitrary size of the cut formula, has already been removed by Theorem 4.1.2. The atomic cut rule,

$$
\mathsf{ai}{\uparrow}\,\frac{S(a,\bar{a})}{S\{\mathsf{f}\}} \quad ,
$$

still is infinitely generating, since there is an infinite choice of atoms. To remove this infinite choice, consider the rule *finitely generating atomic cut*

$$
\mathsf{fai}{\uparrow}\,\frac{S(a,\bar{a})}{S\{\mathsf{f}\}} \quad \text{where } a \text{ or } \bar{a} \text{ appears in the conclusion.}
$$

This rule is finitely generating, and we will show that we can easily transform a proof into one where the only cuts that appear are fai↑ instances.

Take a proof in SKS \ {aw↑}. Single out the bottommost instance of ai↓ that violates the proviso of fai↑:

$$
\mathsf{ai}{\uparrow}\,\frac{\overset{\displaystyle\|}{S(a,\bar{a})}}{S\{\mathsf{f}\}} \quad ,
$$

where neither a nor $\bar{a}$ appears in $S\{\mathsf{f}\}$. We can then replace all instances of a and $\bar{a}$ in the proof above the cut with t and f, respectively, to obtain a proof of $S\{\mathsf{f}\}$. It is easy to check that all rule instances stay intact or become instances of the equivalence rule. The cut, for example, is replaced by an instance of the equivalence rule

$$\mathsf{ai}{\uparrow}\,\frac{S(a,\bar{a})}{S\{\mathsf{f}\}} \qquad \rightsquigarrow \qquad =\frac{S(\mathsf{t},\mathsf{f})}{S\{\mathsf{f}\}} \qquad .$$

Please notice that if a or $\bar{a}$ appeared in $S\{\mathsf{f}\}$, then this would not work, because it could destroy the rule instance below $S\{\mathsf{f}\}$.

Proceeding inductively upwards, we remove all infinitely generating atomic cuts.

5.1.2   A Finitely Generating System for Propositional Logic

We now define the finitely generating system FKS to be

$$(\mathsf{SKS} \setminus \{\mathsf{ai}{\uparrow}, \mathsf{aw}{\uparrow}\}) \cup \{\mathsf{fai}{\uparrow}\} \quad ,$$

and, for what we said above, state

Theorem 5.1.1. *Each formula is provable in system* SKS *if and only if it is provable in system* FKS.

To put our finitely generating system at work, we now show consistency of system SKS by showing consistency of system FKS.

Theorem 5.1.2. *The unit* f *is not provable in system* FKS.

Proof. No atoms, but only f, t can appear in such a proof. It is easy to show that f is not equivalent to t. Then we show that no rule can have a premise equivalent to t and a conclusion equivalent to f. This is simply done by inspection of all the rules in FKS.   □

From the two theorems above we immediately get

Corollary 5.1.3. *The unit* f *is not provable in system* SKS.

Here is an example that makes use of the symmetry of the calculus of structures by flipping derivations: assuming that we can not prove f in the system, having a proof of R implies that there is no proof of $\bar{R}$. We assume that we have both proofs:

$$\overset{\Vert}{R} \qquad \text{and} \qquad \overset{\Vert}{\bar{R}} \qquad ,$$

dualise the proof of R, to get

$$\begin{array}{c}\bar{R}\\\parallel\\\mathsf{f}\end{array}\quad,$$

and compose this derivation with the proof of $\bar{R}$ to get a proof of f, which is a contradiction:

$$\begin{array}{c}\parallel\\\bar{R}\\\parallel\\\mathsf{f}\end{array}\quad,$$

so we obtain

Corollary 5.1.4. *If a formula is provable in system* SKS *then its negation is not provable.*

5.2 Predicate Logic

5.2.1 Eliminating Infinite Choice in Inference Rules

There are three infinitely generating rules in system SKSq: the atomic co-weakening, the atomic cut, and the instantiation rule. The atomic co-weakening can be easily removed as in the propositional case. In the following we will see how to replace atomic cut and instantiation by finitely generating rules without affecting provability.

The Atomic Cut Rule

Let us take a closer look at the atoms in the atomic cut rule:

$$\mathsf{ai}{\uparrow}\,\frac{S(p(\tau_1,\ldots,\tau_n),\overline{p(\tau_1,\ldots,\tau_n)})}{S\{\mathsf{f}\}}\quad.$$

There are both an infinite choice of predicate symbols p and an infinite choice of terms for each argument of p. Let $\vec{\tau}$ denote $\tau_1,\ldots,\tau_n$ and $\vec{x}$ denote $x_1,\ldots,x_n$. Since cuts can be applied inside existential quantifiers, we can delegate the choice of terms to a sequence of $\mathsf{n}{\downarrow}$ instances:

$$\mathsf{ai}{\uparrow}\,\frac{S(p(\vec{\tau}),\overline{p(\vec{\tau})})}{S\{\mathsf{f}\}}\quad\leadsto\quad\mathsf{n}{\downarrow}^n\,\frac{\dfrac{S(p(\vec{\tau}),\overline{p(\vec{\tau})})}{\mathsf{ai}{\uparrow}\,\dfrac{S\{\exists\,\vec{x}\;(p(\vec{x}),\overline{p(\vec{x})})\}}{=\dfrac{S\{\exists\,\vec{x}\;\mathsf{f}\}}{S\{\mathsf{f}\}}}}}{}\quad.$$

The remaining cuts are restricted in that they do not introduce arbitrary terms but just existential variables. Let us call this restricted form vai↑:

$$\text{vai}{\uparrow}\ \frac{S(p(\overrightarrow{x}),\overline{p(\overrightarrow{x})})}{S\{\mathsf{f}\}}\quad .$$

The only infinite choice that remains is the one of the predicate symbol p. To remove it, consider the rule *finitely generating atomic cut*

$$\text{fai}{\uparrow}\ \frac{S(p(\overrightarrow{x}),\overline{p(\overrightarrow{x})})}{S\{\mathsf{f}\}}\quad\text{where } p \text{ appears in the conclusion.}$$

This rule is finitely generating, and we will show that we can easily transform a proof into one where the only cuts that appear are fai↑ instances.

Take a proof in the system we obtained so far, that is SKSq without aw↑, and with vai↑ instead of ai↑. Single out the bottommost instance of vai↓ that violates the proviso of fai↑:

$$\text{vai}{\uparrow}\ \frac{\overset{\Vert}{\rule{0pt}{0pt}}\ \overline{S(p(\overrightarrow{x}),\overline{p(\overrightarrow{x})})}}{S\{\mathsf{f}\}}\quad ,$$

where p does not appear in $S\{\mathsf{f}\}$. We can then replace all instances of $p(\overrightarrow{x})$ and $\overline{p(\overrightarrow{x})}$ in the proof above the cut with t and f, respectively, to obtain a proof of $S\{\mathsf{f}\}$. It is easy to check that all rule instances stay intact or become instances of the equivalence rule. As in the propositional case, the cut rule is replaced by an instance of equivalence

$$\text{vai}{\uparrow}\ \frac{S(a,\bar{a})}{S\{\mathsf{f}\}}\quad\rightsquigarrow\quad =\frac{S(\mathsf{t},\mathsf{f})}{S\{\mathsf{f}\}}\quad .$$

Please notice that if p appeared in $S\{\mathsf{f}\}$, then this would not work, because it could destroy the rule instance below $S\{\mathsf{f}\}$.

Proceeding inductively upwards, we remove all infinitely generating atomic cuts.

The Instantiation Rule

The same techniques also work for instantiation. Consider these two restricted versions of n↓:

$$\text{n}{\downarrow}_1\ \frac{S\{R[x/f(\overrightarrow{x})]\}}{S\{\exists x R\}}\quad\text{and}\quad\text{n}{\downarrow}_2\ \frac{S\{R[x/y]\}}{S\{\exists x R\}}\quad .$$

An instance of $n\downarrow$ that is not an instance of $n\downarrow_2$ can inductively be replaced by instances of $n\downarrow_1$ (choose variables for $\vec{x}$ that are not free in R):

$$n\downarrow \frac{S\{R[x/f(\vec{\tau})]\}}{S\{\exists x R\}} \quad\leadsto\quad \begin{array}{c} n\downarrow^n \dfrac{S\{R[x/f(\vec{\tau})]\}}{S\{\exists\ \vec{x}\ R[x/f(\vec{x})]\}} \\[2mm] n\downarrow_1 \dfrac{}{\dfrac{S\{\exists\ \vec{x}\ \exists x R\}}{=\overline{S\{\exists x R\}}}} \end{array} \quad .$$

This process can be repeated until all instances of $n\downarrow$ are either instances of $n\downarrow_1$ or $n\downarrow_2$.

Now consider the following finitely generating rules,

$$\mathsf{fn}\downarrow_1 \frac{S\{R[x/f(\vec{x})]\}}{S\{\exists x R\}} \quad\text{and}\quad \mathsf{fn}\downarrow_2 \frac{S\{R[x/y]\}}{S\{\exists x R\}} \quad .$$

where $\mathsf{fn}\downarrow_1$ carries the proviso that the function symbol f either occurs in the conclusion or is a fixed constant c, and $\mathsf{fn}\downarrow_2$ carries the proviso that the variable y appears in the conclusion (no matter whether free or bound or in a vacuous quantifier).

Infinitely generating instances of $n\downarrow_1$ and $n\downarrow_2$, i.e. those that are not instances of $\mathsf{fn}\downarrow_1$ and $\mathsf{fn}\downarrow_2$, respectively, are easily replaced by finitely generating rules similarly to how the infinitely generating cuts were eliminated. Take the constant symbol c that is fixed in the proviso of $\mathsf{fn}\downarrow_1$, and throughout the proof above an infinitely generating instance of $n\downarrow_1$, replace all terms that are instances of $f(\vec{x})$ by c. For $n\downarrow_2$ we do the same to all occurrences of y, turning it into an instance of $\mathsf{fn}\downarrow_1$.

The Equivalence and Co-instantiation Rules

The equivalence rule can be broken up into several rules, just like in the propositional case. Those rules are finitely generating except for variable renaming and vacuous quantifier, which, technically speaking, have an infinite choice of names for bound variables. The same goes for the co-instantiation rule. Of course these rules can be made finitely generating since the choice of names of bound variables does not matter. There is nothing deep in it: the only reason for me to tediously show this obvious fact is to avoid giving the impression that I hide infinity under the carpet of the equivalence. The need for the argument below just comes from a syntax which has infinitely many different objects for essentially the same thing, e.g. $\forall x p(x), \forall y p(y)$ and $\forall y \forall x p(x) \ldots$. If you are not concerned about this 'infinite' choice of names of bound variables, then please feel invited to skip ahead to the next section.

Consider the following rules for variable renaming and vacuous quantifier, they all carry the proviso that x is not free in R:

$$\alpha\downarrow \frac{S\{\forall x R[y/x]\}}{S\{\forall y R\}} \qquad \alpha\uparrow \frac{S\{\exists x R[y/x]\}}{S\{\exists y R\}}$$

$$\mathsf{v}\downarrow \frac{S\{\exists x R\}}{S\{R\}} \qquad \mathsf{v}\uparrow \frac{S\{R\}}{S\{\forall x R\}}$$

Let us now consider proofs in the system that is obtained from SKSq by restricting the equivalence rule to not include vacuous quantifier and variable renaming and by adding the above rules. This system is strongly equivalent to SKSq as can easily be checked.

The rule $\mathsf{v}\uparrow$ is clearly finitely generating. Let us see how to replace the infinitely generating rule $\mathsf{v}\downarrow$ by finitely generating rules, the same technique also works for the rules $\alpha\uparrow$ and $\alpha\downarrow$. Consider the finitely generating rule $\mathsf{fv}\downarrow_1$, which is $\mathsf{v}\downarrow$ with the added proviso that x occurs in the conclusion (no matter whether bound or free or in a vacuous quantifier) and the infinitely generating rule $\mathsf{v}\downarrow'$ which is $\mathsf{v}\downarrow$ with the proviso that x does not occur in the conclusion.

Fix a total order on variables. Let $\mathsf{fv}\downarrow_2$ be $\mathsf{v}\downarrow$ with the proviso that x is the lowest variable in the order that does not occur in the conclusion. This rule is clearly finitely generating: there is no choice.

Each instance of $\mathsf{v}\downarrow$ is either an instance of $\mathsf{fv}\downarrow_1$ or of $\mathsf{v}\downarrow'$. In a given proof, all instances of $\mathsf{v}\downarrow'$ can be replaced by instances of $\mathsf{fv}\downarrow_2$ as follows, as we we will see now. Starting from the conclusion, going up in the proof, identify the first infinitely generating vacuous quantifier rule:

$$\mathsf{v}\downarrow' \frac{\overline{\overline{}} \atop S\{\exists x R\}}{\underset{T}{\overset{\|}{S\{R\}}}} \qquad x \text{ does not occur in } S\{R\} \qquad ,$$

where x is not the lowest in our fixed order that does not occur in the conclusion. Let y be the lowest variable that does not occur in the conclusion. Now, throughout the proof above, do the following:

1. Choose a variable z that does not occur in the proof. Replace y by z.

2. Replace x by y.

By definition neither x nor y occur in the conclusion, so the conclusion is not broken. All the replacements respect that variable occurrences with different names stay different and variable occurrences with the same names stay the same. So the proof above stays intact. Replace the $\mathsf{v}{\downarrow}'$ instance by a $\mathsf{fv}{\downarrow}_2$ instance and proceed inductively upwards.

5.2.2 A Finitely Generating System for Predicate Logic

We now obtain the finitely generating system FKSq from SKSq by removing the rules $\{\mathsf{ai}{\uparrow}, \mathsf{aw}{\uparrow}, \mathsf{n}{\downarrow}\}$ and adding the rules shown in Figure 5.1. Note that, strictly speaking, this system is not finitely generating because of the infinite choice in naming bound variables. As we have seen in the last section, we can easily obtain a system that is finitely generating. However, that would be a bit pedantic and would clutter up what is relevant here: the finite choice in the cut and in the instantiation rule.

$$
\mathsf{fai}{\uparrow}\ \frac{S(p(\vec{x}), \overline{p(\vec{x})})}{S\{\mathsf{f}\}} \qquad \text{where } p \text{ appears in the conclusion}
$$

$$
\mathsf{fn}{\downarrow}_1\ \frac{S\{R[x/f(\vec{x})]\}}{S\{\exists x R\}} \qquad \begin{array}{l}\text{where } f \text{ either occurs in the con-}\\ \text{clusion or is a fixed constant } c\end{array}
$$

$$
\mathsf{fn}{\downarrow}_2\ \frac{S\{R[x/y]\}}{S\{\exists x R\}} \qquad \text{where } y \text{ appears in the conclusion}
$$

Figure 5.1: Finitely generating rules

From the previous sections we know that the following theorem holds:

Theorem 5.2.1. *Each formula is provable in system* SKSq *if and only if it is provable in system* FKSq.

Consistency of system FKSq can now be shown just like in the propositional case. Of course, for this purpose it suffices to have finitely generating cut. Having infinite choice in instantiation would not affect the argument.

Theorem 5.2.2. *The unit* f *is not provable in system* FKSq.

From the two theorems above we immediately get

Corollary 5.2.3. *The unit* f *is not provable in system* SKSq.

Just like in the propositional case, we can flip derivations to prove

Corollary 5.2.4. *If a formula is provable in system* SKSq *then its negation is not provable.*

Chapter Summary

Starting from the local deductive systems that were introduced in the previous chapter, we have obtained equivalent systems in which all rules are finitely generating.

Some of the techniques used, for example the replacement of an atom and its dual by t and f, are folklore. However, in order to produce a finitely generating system they have to be combined with the reduction of the cut rule to its atomic form and the ability to apply rules inside the scope of an existential quantifier, which are not available in the sequent calculus.

In the sequent calculus, finite choice can not be achieved by such simple techniques: one relies on cut admissibility. Deep inference and top-down symmetry allow us to obtain finite choice without having to rely on cut admissibility.

Deep and symmetric systems for various modal logics [39], linear logic [41, 42] and various extensions of it [22, 25, 9] and for noncommutative logics [12] are all similar to systems and SKS in the sense that they include rules which follow a scheme [21], which ensures atomicity of cut and identity. So it is certainly possible to use the methods presented here for these logics.

Chapter 6

Cut Elimination

In this chapter we will see a syntactic cut elimination procedure for system SKS. We already know that the cut rule is admissible, both by the semantic argument and by the translations to the sequent calculus that we have seen in Chapter 2. However, it is interesting to study techniques for eliminating the cut without relying on semantics and without relying on translations to formalisms which already have a cut elimination result. Subjects of interest are for example the computational interpretation and the complexity of cut elimination procedures.

To eliminate the cut in system SKS we just have to consider atomic cuts because the general cut rule easily reduces to its atomic form, as we have seen in Chapter 4. This allows for a very simple cut elimination procedure: one induction measure, the cut-rank, just disappears. In fact, the atomicity of the cut formula allows to simply plug-in proofs during the cut elimination process, which is more similar to normalisation in natural deduction than to cut elimination in the sequent calculus.

6.1 Motivation

The two well-known connections between proof theory and language design, *proof search as computation* and *proof normalisation as computation*, have mainly used different proof-theoretic formalisms. While designers of functional programming languages prefer natural deduction, because of the close correspondence between proof normalisation and reduction in related term calculi [15, 30], designers of logic programming languages prefer the sequent calculus [29], because infinite choice and much of the unwanted nondeterminism is limited to the cut rule, which can be eliminated.

System SKS has an explicit cut rule, which is admissible. Thus, in principle, it is as suitable for proof search as systems in the sequent calculus. System

SKS also admits a cut elimination procedure which is similar to normalisation in natural deduction. It could thus allow us to develop both the proof search and the proof normalisation paradigm of computation in the same formalism and starting from the same system of rules.

6.2   Cut Elimination in Sequent Calculus and Natural Deduction

Cut elimination in the sequent calculus and normalisation in natural deduction, widely perceived as 'morally the same', differ quite a bit, technically. Compared to cut elimination, (weak) normalisation is simpler, involving neither permutation of a multicut rule, nor induction on the cut-rank. The equivalent of a cut in natural deduction, a succession of an introduction and an elimination rule, is eliminated as shown in Figure 6.1: first, assumption A and all its copies are removed from Δ_1, as indicated by the arrow in the left. Second, the derivation Δ_2, with the context strengthened accordingly, is plugged into all the leaves of Δ_1 where assumption A was used. This plugging is indicated by the arrows on the right.

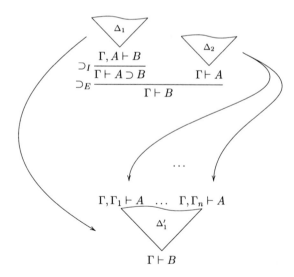

Figure 6.1: Normalisation in natural deduction

This method relies on the fact that no rule inside Δ_1 can change the premise A, which is why it does not work for the sequent calculus. To eliminate a cut in the sequent calculus, one has to cope with the fact that rules may be applied to both, the cut formula and its dual. This requires to permute up the cut rule step-by-step in a complex procedure. Figure 6.2 shows how a paradigmatic example of such a step. To see that such a procedure terminates one has to keep track not only of the distance from the cut to the top of the proof, but also of the size of the cut formula.

However, given a cut with an *atomic* cut formula a inside a sequent calculus proof, we can trace the occurrence of a and its copies produced by contraction, identify all the leaves where they are used in identity axioms, and plug in subproofs in very much the same way as in natural deduction. The problem for the sequent calculus is that cuts are not atomic, in general.

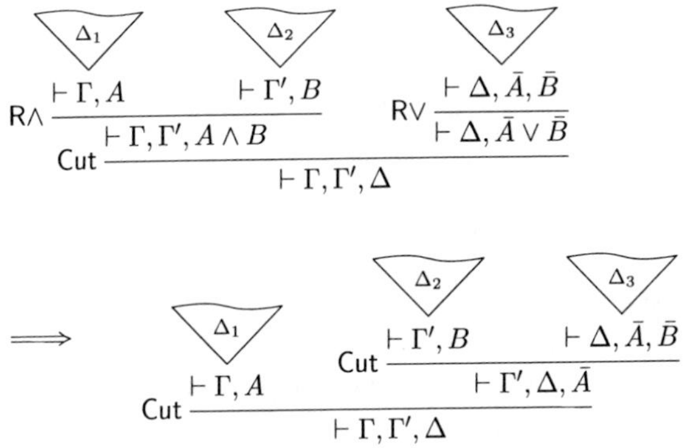

Figure 6.2: Cut elimination in the sequent calculus

6.3 Cut Elimination in the Calculus of Structures

In the calculus of structures, there is more freedom in applying inference rules than in the sequent calculus. While this allows for a richer combinatorial analysis of proofs, it is a significant challenge for cut elimination. During cut elimination, the sequent calculus allows to get into the crucial situation where on one branch a logical rule applies to the main connective of the cut formula and on the other branch the corresponding rule applies to the dual connective of the dual cut formula. In the calculus of structures, rules apply deep inside a context, they are not restricted to main connectives. The methodology of the sequent calculus thus does not apply to the calculus of structures. For example, one cannot permute the cut over the switch rule. One can generalise the cut in order to permute it over switch, but this requires a case analysis that is far more complicated than in the sequent calculus. Contraction is an even bigger problem. Despite many efforts, no cut elimination procedure along these lines has been found for system SKS.

Two new techniques were developed to eliminate cuts in the calculus of structures. The first is called *decomposition*, and has been used in [24, 42] for some systems related to linear logic. Proving termination of decomposition is rather involved [42]. It makes essential use of the exponentials of linear logic which restrict the use of contraction. So far, this technique could not be used for classical logic with its unrestricted contraction. The second technique is called *splitting* [22], and essentially makes available a situation corresponding to the one described above for the sequent calculus. Splitting covers the broadest range of systems in the calculus of structures, it not only applies to the systems mentioned above, but has recently also been applied to system SKS (but there are no references yet). Compared to splitting, the procedure given here is much simpler. In fact, I do not know of any other system with such a simple cut elimination procedure.

6.4   The Cut Elimination Procedure

The cut elimination procedure presented here is based on the fact that the cut rule is already reduced to atomic form, which can be taken for granted as we have seen in Chapter 4. This allows to plug-in proofs as in normalisation in natural deduction.

There is one difference, however. In the sequent calculus as well as in sequent-style natural deduction, a derivation is a tree. Seen bottom-up, a cut splits the tree into two branches. To apply a cut, one is forced to split the context among the two branches (in the case of multiplicative context treatment) or to duplicate the context (in the case of additive context treatment). In the calculus of structures, the cut rule does not split the proof. This is convenient for proof construction, because one is not forced to split or duplicate a context in order to apply the cut rule. However, it is an obstacle in the way of getting a cut elimination procedure.

To overcome this obstacle, we split the proof during cut elimination. Apart from that, the procedure is similar to normalisation in natural deduction. The idea is illustrated in Figure 6.3: we duplicate the proof above a cut and remove atom a from one copy, illustrated by the arrow to the right, and remove the atom $\bar{a}$ from the other copy, illustrated by the arrow to the left. We choose one of the obtained proofs, the one on the left in this case, and replace a by R throughout the proof, illustrated by the arrow on the left. This process breaks some instances of identity, for that reason the obtained object is not a proof. These instances then are fixed by substituting a copy

of the the proof on the right for every broken instance, as illustrated by the
arrows on the right. A contraction is applied at the bottom to obtain a
cut-free proof of R.

At this point it may be helpful to compare Figure 6.3 with Figure 6.1. A
more formal presentation of the cut elimination procedure follows below.

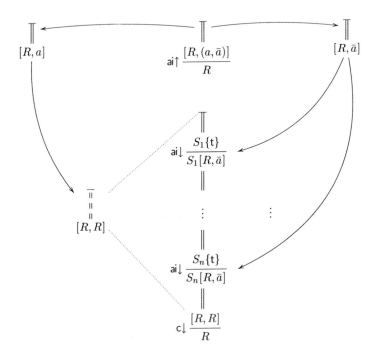

Figure 6.3: Elimination of one atomic cut

In contrast to the sequent calculus, the cut is not the only problematic rule
in system SKS. The rule aw↑ also induces infinite choice in proof-search.
Fortunately, we can not only eliminate the cut rule, but also the other up-
rules. Each up-rule individually can be shown to be admissible for system
KS. However, since we are going to eliminate the cut anyway, to eliminate
rules aw↑ and ac↑ the following lemma is sufficient.

Lemma 6.4.1. *Each rule in* SKS *is derivable for identity, cut, switch and its
dual rule.*

Proof. An instance of $\rho\uparrow\dfrac{S\{T\}}{S\{R\}}$ can be replaced by

$$
\rho\downarrow\ \underset{\mathsf{i}\uparrow}{\underset{\displaystyle}{\cfrac{\mathsf{i}\downarrow\ \cfrac{\mathsf{s}\ \cfrac{S\{T\}}{S(T,[R,\bar{R}])}}{S[R,(T,\bar{R})]}}{S[R,(T,\bar{T})]}}}\quad.
$$

The same holds for down-rules.

□

The atomic cut rule, which applies arbitrarily deep inside a formula, can easily be replaced by a more restricted version of it. It will then suffice to eliminate this restricted version in order to eliminate all cuts.

Definition 6.4.2. An instance of atomic cut is called *shallow atomic cut* if it is of the following form:

$$
\mathsf{ai}\uparrow\ \frac{[S,(a,\bar{a})]}{S}\quad.
$$

Lemma 6.4.3. *The atomic cut is derivable for shallow atomic cut and switch.*

Proof. An easy induction locally replaces an instance of atomic cut by a shallow atomic cut followed by instances of switch. The relevant transformation is

$$
\mathsf{ai}\uparrow\ \frac{S([R,(a,\bar{a})],T)}{=\ \dfrac{S([R,\mathsf{f}],T)}{S(R,T)}}
\qquad\rightsquigarrow\qquad
\mathsf{ai}\uparrow\ \frac{\mathsf{s}\ \dfrac{S[(R,(a,\bar{a})),T]}{S[(R,T),(a,\bar{a})]}}{=\ \dfrac{S[(R,T),\mathsf{f}]}{S(R,T)}}\quad.
$$

□

During cut elimination in the sequent calculus one has access to two proofs above the cut such that the cut formula is in the conclusion of one proof and the dual of the cut formula is in the conclusion of the other proof. In the calculus of structures, we just have one proof above the cut and its conclusion contains both, the (atomic) cut formula and its dual. To gain access to two proofs as in the sequent calculus, we need the following lemma.

Lemma 6.4.4. *Each proof* $\left\|\begin{matrix}\mathsf{KS}\\T\{a\}\end{matrix}\right.$ *can be transformed into a proof* $\left\|\begin{matrix}\mathsf{KS}\\T\{\mathsf{t}\}\end{matrix}\right.$.

Proof. Starting with the conclusion, going up in the proof, in each formula we replace the occurrence of a and its copies, that are produced by contractions, by the unit t. Replacements inside the context of any rule instance leave this rule instance intact. Instances of the rules m and s remain intact, also in the case that atom occurrences are replaced inside redex and contractum. Instances of the other rules are replaced by the following derivations:

$$\mathsf{ac}{\downarrow} \frac{S[a,a]}{S\{a\}} \quad \rightsquigarrow \quad = \frac{S[\mathsf{t},\mathsf{t}]}{S\{\mathsf{t}\}}$$

$$\mathsf{aw}{\downarrow} \frac{S\{\mathsf{f}\}}{S\{a\}} \quad \rightsquigarrow \quad = \frac{\dfrac{S\{\mathsf{f}\}}{S([\mathsf{t},\mathsf{t}],\mathsf{f})}}{\mathsf{s} \dfrac{S[\mathsf{t},(\mathsf{t},\mathsf{f})]}{S\{\mathsf{t}\}}}$$

$$\mathsf{ai}{\downarrow} \frac{S\{\mathsf{t}\}}{S[a,\bar{a}]} \quad \rightsquigarrow \quad = \frac{\dfrac{S\{\mathsf{t}\}}{S[\mathsf{t},\mathsf{f}]}}{\mathsf{aw}{\downarrow} \dfrac{S[\mathsf{t},\mathsf{f}]}{S[\mathsf{t},\bar{a}]}} \quad .$$

$\square$

Properly equipped, we now turn to cut elimination.

Theorem 6.4.5. *Each proof* $\overset{\mathsf{SKS}}{\underset{T}{\big\|}}$ *can be transformed into a proof* $\overset{\mathsf{KS}}{\underset{T}{\big\|}}$.

Proof. By Lemma 6.4.1, the only rule left to eliminate is the cut. By Lemma 6.4.3, we replace all cuts by shallow cuts. The topmost instance of cut, together with the proof above it, is singled out:

$$\overset{\mathsf{KS}\cup\{\mathsf{ai}{\uparrow}\}}{\underset{T}{\Big\|}} \quad = \quad \mathsf{ai}{\uparrow} \frac{\overset{\Pi \big\| \mathsf{KS}}{[R,(a,\bar{a})]}}{\dfrac{R}{\underset{T}{\Delta \big\| \mathsf{KS}\cup\{\mathsf{ai}{\uparrow}\}}}} \quad .$$

Lemma 6.4.4 is applied twice on Π to obtain

$$\overset{\Pi_1 \big\| \mathsf{KS}}{[R,a]} \quad \text{and} \quad \overset{\Pi_2 \big\| \mathsf{KS}}{[R,\bar{a}]} \quad .$$

Starting with the conclusion, going up in proof Π_1, in each formula we replace the occurrence of a and its copies, that are produced by contractions, by the formula R.

Replacements inside the context of any rule instance leave the rule instance intact. Instances of the rules m and s remain intact, also in the case that atom occurrences are replaced inside redex and contractum. Instances of ac$\downarrow$ and aw$\downarrow$ are replaced by their general versions:

$$\mathsf{ac}{\downarrow}\,\frac{S[a,a]}{S\{a\}} \quad \rightsquigarrow \quad \mathsf{c}{\downarrow}\,\frac{S[R,R]}{S\{R\}}$$

$$\mathsf{aw}{\downarrow}\,\frac{S\{\mathsf{f}\}}{S\{a\}} \quad \rightsquigarrow \quad \mathsf{w}{\downarrow}\,\frac{S\{\mathsf{f}\}}{S\{R\}} \quad .$$

Instances of ai$\downarrow$ are replaced by $S\{\Pi_2\}$:

$$\mathsf{ai}{\downarrow}\,\frac{S\{\mathsf{t}\}}{S[a,\bar a]} \quad \rightsquigarrow \quad
\begin{array}{c} S\{\mathsf{t}\} \\ S\{\Pi_2\}\Big\|\mathsf{KS} \\ S[R,\bar a] \end{array} \quad .$$

The result of this process of substituting Π_2 into Π_1 is a proof Π_3, from which we build

$$\begin{array}{c}
\Pi_3\Big\|\mathsf{KS} \\
\mathsf{c}{\downarrow}\,\dfrac{[R,R]}{R} \\
\Delta\Big\|\mathsf{KS}\cup\{\mathsf{ai}\downarrow\} \\
T
\end{array}$$

Proceed inductively downward with the remaining instances of cut. $\qquad\square$

Chapter Summary

We have seen a very simple cut elimination procedure for system SKS. Its simplicity comes from the fact that the cut is already reduced to atomic form, which allows to use the technique of plugging proofs, similar to what one does in normalisation for natural deduction.

Chapter 7

Normal Forms

Derivations can be arranged into consecutive phases such that each phase uses only certain rules. We call this property *decomposition*. Decomposition theorems thus provide normal forms for derivations. A classic example of a decomposition theorem in the sequent calculus is Gentzen's Mid-Sequent Theorem [16], which states the possibility of decomposing a cut-free proof into a lower phase with contraction and quantifier rules and an upper phase with propositional rules only. In this chapter we will first see the technique of permuting rules. Then three decomposition theorems are presented which state the possibility of pushing all instances of a certain rule to the top and all instances of its dual rule to the bottom of a derivation. Except for the first, these decomposition theorems do not have analogues in the sequent calculus. The three decomposition theorems will then be combined to yield a decomposition of derivations into seven phases where the systems used in each phase are pairwise disjoint. We then see how these results extend to predicate logic and why they do not exist in the sequent calculus. In the last section of the chapter we see a decomposition theorem for propositional proofs which generalises cut elimination as well as Craig interpolation [10].

7.1 Permutation of Rules

Informally, permuting an instance of a rule over an instance of another rule in a derivation means exchanging the two rules without breaking the derivation.

Definition 7.1.1. A rule ρ *permutes over* a rule π (or π *permutes under* ρ)

if for every derivation $\pi \dfrac{\begin{array}{c} T \\ \hline U \end{array}}{R}\rho$ there is a derivation $\rho \dfrac{\begin{array}{c} T \\ \hline V \end{array}}{R}\pi$ for some formula V.

Since deep inference grants more freedom in applying inference rules than shallow inference of the sequent calculus, it also gives rise to more permutability.

Lemma 7.1.2. *The rule* $\mathsf{ac}{\downarrow}$ *permutes under the rules* $\mathsf{aw}{\downarrow}$, $\mathsf{ai}{\downarrow}$, s *and* m. *Dually, the rule* $\mathsf{ac}{\uparrow}$ *permutes over the rules* $\mathsf{aw}{\uparrow}$, $\mathsf{ai}{\uparrow}$, s *and* m.

Proof. Given an instance of $\mathsf{ac}{\downarrow}$ above an instance of a rule $\rho \in \{\mathsf{aw}{\downarrow}, \mathsf{ai}{\downarrow}, \mathsf{s}, \mathsf{m}\}$, the redex of $\mathsf{ac}{\downarrow}$ can be a subformula of the context of ρ. Then we permute as follows:

$$
\mathsf{ac}{\downarrow} \frac{S'\{U\}}{\rho \dfrac{S\{U\}}{S\{R\}}}
\qquad \rightsquigarrow \qquad
\rho \frac{S'\{U\}}{\mathsf{ac}{\downarrow} \dfrac{S'\{R\}}{S\{R\}}} \quad .
$$

Since the redex of $\mathsf{ac}{\downarrow}$ is an atom, the only other possibility occurs in case that ρ is s or m: the redex of $\mathsf{ac}{\downarrow}$ can be a subformula of the contractum of ρ. Then we permute as in the following example of a switch rule, where $T\{\ \}$ is a formula context:

$$
\mathsf{ac}{\downarrow} \frac{S([R, T[a, a]], U)}{\mathsf{s} \dfrac{S([R, T\{a\}], U)}{S[R, (T\{a\}, U)]}}
\qquad \rightsquigarrow \qquad
\mathsf{s} \frac{S([R, T[a, a]], U)}{\mathsf{ac}{\downarrow} \dfrac{S([R, T[a, a]], U)}{S[R, (T\{a\}, U)]}} \quad .
$$

(And dually for $\mathsf{ac}{\uparrow}$.)   □

Lemma 7.1.3. *The rule* $\mathsf{aw}{\downarrow}$ *permutes under the rules* $\mathsf{ai}{\downarrow}$, s *and* m. *Dually, the rule* $\mathsf{aw}{\uparrow}$ *permutes over the rules* $\mathsf{ai}{\uparrow}$, s *and* m.

Proof. Similar to the proof of Lemma 7.1.2.   □

We now turn to the decomposition results.

7.2   Separating Identity and Cut

Given that in system SKS identity is a rule and not an axiom as in the sequent calculus, a natural question to ask is whether the applications of the identity rule can be restricted to the top of a derivation. For proofs, this question is already answered positively by the semantic cut elimination argument, Theorem 2.4.8 on page 22. It turns out that it is also true for derivations in general. Because of the duality between $\mathsf{ai}{\downarrow}$ and $\mathsf{ai}{\uparrow}$ we can also push the cuts to the bottom of a derivation. While this can be obtained in the sequent calculus (by using cut elimination), it can not be done with a simple permutation argument as we do.

The following rules are called *super switch* and *super co-switch*:

$$\mathsf{ss}{\downarrow}\;\frac{S\{T\{R\}\}}{S[R,T\{f\}]} \qquad \text{and} \qquad \mathsf{ss}{\uparrow}\;\frac{S(R,T\{\mathsf{t}\})}{S\{T\{R\}\}} \quad.$$

Lemma 7.2.1. *The rule $\mathsf{ss}{\downarrow}$ is derivable for $\{\mathsf{s}\}$. Dually, the rule $\mathsf{ss}{\uparrow}$ is derivable for $\{\mathsf{s}\}$.*

Proof. Given an instance

$$\mathsf{ss}{\uparrow}\;\frac{S(R,T\{\mathsf{t}\})}{S\{T\{R\}\}} \quad,$$

we show by structural induction on $T\{\ \}$ how to replace it by switches. The proof for $\mathsf{ss}{\downarrow}$ is dual.

1. $T\{\ \}$ is empty. Then the given instance can be replaced by an instance of the equivalence rule.

2. $T\{\ \} = [U, V\{\ \}]$. Apply the induction hypothesis on

$$\mathsf{ss}{\uparrow}\;\frac{\mathsf{s}\;\dfrac{S(R,[U,V\{\mathsf{t}\}])}{S[U,(R,V\{\mathsf{t}\})]}}{S[U,V\{R\}]} \quad.$$

3. $T\{\ \} = (U, V\{\ \})$. Apply the induction hypothesis on

$$\mathsf{ss}{\uparrow}\;\frac{=\;\dfrac{S(R,(U,V\{\mathsf{t}\}))}{S(U,R,V\{\mathsf{t}\})}}{S(U,V\{R\})} \quad.$$

$\square$

We have already seen the shallow atomic cut in the previous chapter. Now we also consider a shallow identity and we suffix an s to the name of the rule to denote that it is shallow:

$$\mathsf{ai_s}{\downarrow}\;\frac{S}{(S,[a,\bar{a}])} \qquad \text{and} \qquad \mathsf{ai_s}{\uparrow}\;\frac{[S,(a,\bar{a})]}{S}$$

are called *shallow atomic identity* and *shallow atomic cut*, respectively.

Lemma 7.2.2. *The rule $\mathsf{ai}{\downarrow}$ is derivable for $\{\mathsf{ai_s}{\downarrow},\mathsf{s}\}$. Dually, the rule $\mathsf{ai}{\uparrow}$ is derivable for $\{\mathsf{ai_s}{\uparrow},\mathsf{s}\}$.*

Proof. An instance of $\mathsf{ai}{\downarrow}$ can be replaced by an instance of $\mathsf{ai_s}{\downarrow}$ followed by an instance of $\mathsf{ss}{\uparrow}$, as follows:

$$\mathsf{ai_s}{\downarrow}\,\frac{S\{\mathsf{t}\}}{\underset{\mathsf{ss}{\uparrow}}{\;\dfrac{(S\{\mathsf{t}\},[a,\bar{a}])}{S[a,\bar{a}]}}}\quad.$$

The rule $\mathsf{ss}{\uparrow}$ can in turn be replaced by a derivation of switches by Lemma 7.2.1. (And dually for $\mathsf{ai}{\uparrow}$.)   □

Theorem 7.2.3.

$$\text{For every derivation } \begin{array}{c} T \\ \Big\| \mathsf{SKS} \\ R \end{array} \text{ there is a derivation } \begin{array}{c} T \\ \|\{\mathsf{ai}{\downarrow}\} \\ V \\ \|\mathsf{SKS}\backslash\{\mathsf{ai}{\downarrow},\mathsf{ai}{\uparrow}\} \\ U \\ \|\{\mathsf{ai}{\uparrow}\} \\ R \end{array}.$$

Proof. We replace atomic identities by shallow atomic identities and switches using Lemma 7.2.2 and do the same for the cuts. The rule $\mathsf{ai_s}{\downarrow}$ trivially permutes over every rule in SKS, since its premise is just a schematic formula. Dually, the rule $\mathsf{ai_s}{\uparrow}$ trivially permutes under every rule in SKS. Instances of $\mathsf{ai_s}{\downarrow}$ and $\mathsf{ai_s}{\uparrow}$ are instances of $\mathsf{ai}{\downarrow}$ and $\mathsf{ai}{\uparrow}$, respectively.   □

7.3   Separating Contraction

Contraction allows the repeated use of a formula in a proof by allowing us to copy it at will. It should be possible to copy everything needed in the beginning, and then go on with the proof without ever having to copy again. This intuition is made precise by the following theorem and holds for system SKS. There is no such result for the sequent calculus as we will see in 7.7. There are sequent systems for classical propositional logic that do not have an explicit contraction rule, however, since they involve context sharing, contraction is built into the logical rules and is used throughout the proof.

Theorem 7.3.1.

$$\text{For every proof } \begin{array}{c} \Big\| \mathsf{KS} \\ S \end{array} \text{ there is a proof } \begin{array}{c} \Big\| \mathsf{KS}\backslash\{\mathsf{ac}{\downarrow}\} \\ S' \\ \|\{\mathsf{ac}{\downarrow}\} \\ S \end{array}.$$

Proof. Using Lemma 7.1.2, permute down all instances of $\mathsf{ac}{\downarrow}$, starting with the bottommost.   □

This result is extended to the symmetric system as follows:

Theorem 7.3.2.

For every derivation $\begin{array}{c} T \\ \| \text{SKS} \\ R \end{array}$ *there is a derivation* $\begin{array}{c} T \\ \|\{\text{ac}\uparrow\} \\ V \\ \|\text{SKS}\setminus\{\text{ac}\downarrow,\text{ac}\uparrow\} \\ U \\ \|\{\text{ac}\downarrow\} \\ R \end{array}$.

Proof. Consider the following derivations that can be obtained:

$$
\begin{array}{c}
T \\ \|\text{SKS} \\ R
\end{array}
\overset{1}{\rightsquigarrow}
\begin{array}{c}
\text{i}\downarrow \dfrac{t}{[\bar{T},T]} \\[2pt]
\Big\| \text{SKS} \\[2pt]
[\bar{T},R]
\end{array}
\overset{2}{\rightsquigarrow}
\begin{array}{c}
t \\ \Big\| \text{KS} \\ [\bar{T},R]
\end{array}
\overset{3}{\rightsquigarrow}
\begin{array}{c}
(T,t) \\ \Big\| \text{KS} \\[2pt]
\text{s}\dfrac{(T,[\bar{T},R])}{\text{i}\uparrow\dfrac{[R,(T,\bar{T})]}{R}}
\end{array}
\overset{4}{\rightsquigarrow}
\begin{array}{c}
(T,t) \\ \Big\| \text{KS}\setminus\{\text{ac}\downarrow\} \\
[R',(T,\bar{T}')] \\ \Big\|\{\text{ac}\downarrow\} \\
\text{i}\uparrow\dfrac{[R',(T,\bar{T})]}{R'} \\[2pt]
\Big\|\{\text{ac}\downarrow\} \\ R
\end{array}
.
$$

1. Put the derivation into the context $[\bar{T},\{\ \}]$. On top of the resulting derivation, apply an i$\downarrow$ to obtain a proof.

2. Eliminate all cuts.

3. Put the proof into the context $(T,\{\ \})$. At the bottom of the resulting derivation, apply a switch and a cut to obtain a derivation from T to R.

4. All instances of ac$\downarrow$ are permuted down as far as possible by using Lemma 7.1.2. Note that there are just two kinds of ac$\downarrow$ instances: those that duplicate atoms from R and those that duplicate atoms from $\bar{T}$; there are none that duplicate atoms from T. The first kind, starting with the bottom-most instance, can be permuted down all the way to the bottom of the derivation. The second kind, also starting with the bottom-most instance, can be permuted down until they meet the cut.

Now, starting with the bottom-most ac$\downarrow$ that is above the cut, we apply the transformation

$$
\text{i}\uparrow\dfrac{\text{ac}\downarrow\dfrac{S(U\{a\},\bar{U}[\bar{a},\bar{a}])}{S(U\{a\},\bar{U}\{\bar{a}\})}}{S\{\text{f}\}}
\quad\rightsquigarrow\quad
\text{i}\uparrow\dfrac{\text{ac}\uparrow\dfrac{S(U\{a\},\bar{U}[\bar{a},\bar{a}])}{S(U(a,a),\bar{U}[\bar{a},\bar{a}])}}{S\{\text{f}\}}
$$

and permute the resulting instance of ac↑ all the way up to the top of the derivation. This is possible because no rule in the derivation above changes T. Proceed inductively with the remaining instances of ac↓ above the cut. The resulting derivation has the desired shape.   □

7.4   Separating Weakening

In the sequent calculus, one usually can push all the instances of weakening up to the top of the proof or, to the same effect, build weakening into the identity axiom:

$$\overline{A, \Phi \vdash A, \Psi} \quad .$$

The same *lazy* way of applying weakening can be done in system SKS, cf. Theorem 2.4.8. However, while a proof in which all weakenings occur at the top is certainly more 'normal' than a proof in which weakenings are scattered all over, this is hardly an interesting normal form. In system SKS something more interesting can be done: applying weakening in an *eager* way.

Theorem 7.4.1.

For every proof $\overset{\displaystyle \|\,\text{KS}}{S}$ there is a proof $\overset{\displaystyle \|\,\text{KS}\backslash\{\text{aw}\downarrow\}}{\underset{\displaystyle \underset{S}{\|\,\{\text{aw}\downarrow\}}}{S'}}$.

Proof. Permute down all instances of aw↓, starting with the bottommost. This is done by using Lemma 7.1.3 and the following transformation:

$$\begin{array}{c} \text{aw}\downarrow \dfrac{S[a, \mathsf{f}]}{S[a, a]} \\ \text{ac}\downarrow \dfrac{}{S\{a\}} \end{array} \quad \rightsquigarrow \quad = \dfrac{S[a, \mathsf{f}]}{S\{a\}} \quad .$$

□

Weakening loses information: when deducing $a \lor b$ from a, the information that a holds is lost. Given a proof of a certain statement, do the weakenings lose information that we would like to keep? Can we obtain a proof of a stronger statement by removing them? The theorem above gives an affirmative answer to that question: given a proof of S, it exhibits a weakening-free proof of a formula S', from which S trivially follows by weakenings.

Notation 7.4.2. A derivation $\overset{\displaystyle T}{\underset{\displaystyle R}{\|\,\{\rho\}}}$ of length n is denoted by $\rho^n \dfrac{T}{R}$.

The result given above extends from proofs to arbitrary derivations:

Theorem 7.4.3.

$$\text{For every derivation} \quad \begin{array}{c} T \\ \| \text{SKS} \\ R \end{array} \quad \text{there is a derivation} \quad \begin{array}{c} T \\ \| \{\text{aw}\uparrow\} \\ V \\ \| \text{SKS}\backslash\{\text{aw}\downarrow,\text{aw}\uparrow\} \\ U \\ \| \{\text{aw}\downarrow\} \\ R \end{array} \, .$$

Proof. There is an algorithm that produces a derivation of the desired shape. It consists of two procedures: 1) pushing up all instances of $\text{aw}\uparrow$ and 2) pushing down all instances of $\text{aw}\downarrow$. Those two procedures are repeated alternatingly until the derivation has the desired shape. An instance of $\text{aw}\uparrow$ that is pushed up can turn into an instance of $\text{aw}\downarrow$ when meeting an instance of $\text{ai}\downarrow$, and the dual case can also happen.

In the following, the process of pushing up instances of $\text{aw}\uparrow$ is shown, the process of pushing down instances of $\text{aw}\downarrow$ is dual.

An instance of $\text{aw}\uparrow$ is a special case of a derivation consisting of n instances of $\text{aw}\uparrow$ and is moved up as such, starting with the topmost. In addition to the cases treated in Lemma 7.1.3 there are the following cases:

$$\text{aw}\uparrow^n \cfrac{\text{ac}\uparrow \cfrac{S'\{a\}}{S'(a,a)}}{S(\mathsf{t},a)} \quad \rightsquigarrow \quad =\cfrac{\text{aw}\uparrow^{n-1} \cfrac{S'\{a\}}{S\{a\}}}{S(\mathsf{t},a)}$$

$$\text{aw}\uparrow^n \cfrac{\text{ac}\downarrow \cfrac{S'[a,a]}{S'\{a\}}}{S\{\mathsf{t}\}} \quad \rightsquigarrow \quad =\cfrac{\text{aw}\uparrow^{n+1} \cfrac{S'[a,a]}{S[\mathsf{t},\mathsf{t}]}}{S\{\mathsf{t}\}}$$

$$\text{aw}\uparrow^n \cfrac{\text{aw}\downarrow \cfrac{S'\{\mathsf{f}\}}{S'\{a\}}}{S\{\mathsf{t}\}} \quad \rightsquigarrow \quad =\cfrac{\mathsf{s}\cfrac{=\cfrac{\text{aw}\uparrow^{n-1}\cfrac{S'\{\mathsf{f}\}}{S\{\mathsf{f}\}}}{S([\mathsf{t},\mathsf{t}],\mathsf{f})}}{S[\mathsf{t},(\mathsf{t},\mathsf{f})]}}{S\{\mathsf{t}\}}$$

$$
\mathsf{aw}{\uparrow}^n\,\frac{\mathsf{ai}{\downarrow}\,\dfrac{S'\{\mathsf{t}\}}{S'[a,\bar a]}}{S[\mathsf{t},\bar a]} \quad\rightsquigarrow\quad \mathsf{aw}{\downarrow}\,\frac{\mathsf{aw}{\uparrow}^{n-1}\,\dfrac{S'\{\mathsf{t}\}}{\dfrac{S\{\mathsf{t}\}}{S[\mathsf{t},\mathsf{f}]}}}{S[\mathsf{t},\bar a]}
$$

The process of applying these transformation clearly terminates since the length of the derivation above the topmost instance of $\mathsf{aw}{\uparrow}$ decreases with each step. The algorithm of alternatingly applying the two dual procedures terminates as well since each run of a procedure that does not produce the desired shape strictly decreases the combined number of instances of $\mathsf{ai}{\downarrow}$ and $\mathsf{ai}{\uparrow}$.

□

7.5 Separating all Atomic Rules

Decomposition results can be applied consecutively. Here, all rules that deal with atoms, namely $\mathsf{ai}{\downarrow}$, $\mathsf{ac}{\downarrow}$, $\mathsf{aw}{\downarrow}$ and their duals, are separated from the rules that deal with the connectives, namely s and m:

Theorem 7.5.1. *For every derivation* $\begin{array}{c}T\\ \|\mathsf{SKS}\\ R\end{array}$ *there is a derivation* $\begin{array}{c}T\\ \|\{\mathsf{ac}{\uparrow}\}\\ T_1\\ \|\{\mathsf{aw}{\uparrow}\}\\ T_2\\ \|\{\mathsf{ai}{\downarrow}\}\\ T_3\\ \|\{\mathsf{s},\mathsf{m}\}\\ R_3\\ \|\{\mathsf{ai}{\uparrow}\}\\ R_2\\ \|\{\mathsf{aw}{\downarrow}\}\\ R_1\\ \|\{\mathsf{ac}{\downarrow}\}\\ R\end{array}$.

Proof. We first push contractions to the outside, using Theorem 7.3.2. In the contraction-free part of the obtained derivation, we push weakening to the outside, using the procedure from the proof of Theorem 7.4.3, which does not introduce new instances of contraction. In the contraction- and weakening-free part of the resulting derivation we then separate out identity and cut by applying the procedure from the proof of Theorem 7.2.3, which neither introduces new contractions nor weakenings. □

7.6 Predicate Logic

All decomposition results we have seen so far for the propositional system SKS extend to the predicate logic system SKSq in a straightforward way.

As in the propositional case, atomic identity and cut can be reduced to their shallow versions using the super switch rules. In the predicate case the rules shallow atomic identity and shallow atomic cut are as follows:

$$\mathsf{ai_s}{\downarrow}\,\frac{S}{(S,\forall[a,\bar{a}])} \qquad \text{and} \qquad \mathsf{ai_s}{\uparrow}\,\frac{[S,\exists(a,\bar{a})]}{S} \quad,$$

where $\forall$ and $\exists$ denote sequences of quantifiers that universally close $[a,\bar{a}]$ and existentially close $(a,\bar{a})$, respectively.

The super switch rules for predicate logic,

$$\mathsf{ss}{\downarrow}\,\frac{S\{T\{R\}\}}{S[R,T\{\mathsf{f}\}]} \qquad \text{and} \qquad \mathsf{ss}{\uparrow}\,\frac{S(R,T\{\mathsf{t}\})}{S\{T\{R\}\}} \quad,$$

carry a proviso: quantifiers in T do not bind variables that occur freely in R. This is not a restriction because bound variables can always be renamed such that the proviso is fulfilled.

Lemma 7.6.1. *The rule* $\mathsf{ss}{\downarrow}$ *is derivable for* $\{\mathsf{s},\mathsf{n}{\downarrow},\mathsf{u}{\downarrow}\}$. *Dually, the rule* $\mathsf{ss}{\uparrow}$ *is derivable for* $\{\mathsf{s},\mathsf{n}{\uparrow},\mathsf{u}{\uparrow}\}$.

Proof. The proof is an extension of the proof of corresponding lemma for propositional logic (Lemma 7.2.1 on page 67). I just show the two cases that have to be considered in addition to the proof in the propositional case. Note that the proviso on $\mathsf{ss}{\uparrow}$ ensures that x does not occur freely in R.

1. $T\{\ \} = \forall x U\{\ \}$. Apply the induction hypothesis on

$$\mathsf{ss}{\uparrow}\,\frac{\mathsf{n}{\uparrow}\,\dfrac{=\,\dfrac{S(R,\forall x U\{\mathsf{t}\})}{S\{\forall x(R,\forall x U\{\mathsf{t}\})\}}}{S\{\forall x(R,U\{\mathsf{t}\})\}}}{S\{\forall x U\{R\}\}} \quad.$$

2. $T\{\ \} = \exists x U\{\ \}$. Apply the induction hypothesis on

$$\mathsf{ss}{\uparrow}\,\frac{\mathsf{u}{\uparrow}\,\dfrac{=\,\dfrac{S(R,\exists x U\{\mathsf{t}\})}{S(\forall x R,\exists x U\{\mathsf{t}\})}}{S\{\exists x(R,U\{\mathsf{t}\})\}}}{S\{\exists x U\{R\}\}} \quad.$$

$\square$

Lemma 7.6.2. *The rule* $\mathsf{ai}{\downarrow}$ *is derivable for* $\{\mathsf{ai_s}{\downarrow}, \mathsf{s}, \mathsf{n}{\uparrow}, \mathsf{u}{\uparrow}\}$. *Dually, the rule* $\mathsf{ai}{\uparrow}$ *is derivable for* $\{\mathsf{ai_s}{\uparrow}, \mathsf{s}, \mathsf{n}{\downarrow}, \mathsf{u}{\downarrow}\}$.

Proof.

$$
\text{An instance of} \quad \mathsf{ai}{\downarrow}\,\frac{S\{\mathsf{t}\}}{S[a,\bar a]} \quad \text{is replaced by} \quad
\begin{array}{l}
\mathsf{ai_s}{\downarrow}\,\dfrac{S\{\mathsf{t}\}}{(S\{\mathsf{t}\}, \forall[a,\bar a])} \\[4pt]
\mathsf{ss}{\uparrow}\,\dfrac{}{S\{\forall[a,\bar a]\}} \\[4pt]
\mathsf{n}{\uparrow}^n\,\dfrac{}{S[a,\bar a]}
\end{array} \quad,
$$

where n is the number of free variables in a. (And dually for $\mathsf{ai}{\uparrow}$.) $\square$

Theorem 7.6.3 (Decomposition in Predicate Logic). *All decomposition theorems also hold in the case of predicate logic, i.e. with* SKS *replaced by* SKSq *and* KS *replaced by* KSq. *In Theorem 7.5.1,* $\{\mathsf{s}, \mathsf{m}\}$ *has to be extended by the quantifier rules, i.e.* $\{\mathsf{u}{\downarrow}, \mathsf{u}{\uparrow}, \mathsf{l_1}{\downarrow}, \mathsf{l_1}{\uparrow}, \mathsf{l_2}{\downarrow}, \mathsf{l_2}{\uparrow}, \mathsf{n}{\downarrow}, \mathsf{n}{\uparrow}\}$.

Proof. Identity and cut are separated as in the propositional case, using Lemma 7.6.2 instead of Lemma 7.2.2.

Contraction is separated as in the propositional case, using the proof of Theorem 7.3.2. The only difference is in step four, where instances of $\mathsf{ac}{\downarrow}$ have to be permuted under instances of rules from $\mathsf{KSq} \setminus \mathsf{KS}$. None of those rules except for $\mathsf{n}{\downarrow}$ changes atoms, so $\mathsf{ac}{\downarrow}$ trivially permutes under those instances. It also easily permutes under instances of $\mathsf{n}{\downarrow}$:

$$
\mathsf{ac}{\downarrow}\,\frac{
\mathsf{n}{\downarrow}\,\dfrac{S\{R[a,a]\,[x/\tau]\}}{S\{R\{a\}\,[x/\tau]\}}
}{S\{\exists x R\{a\}\}}
\quad \rightsquigarrow \quad
\mathsf{n}{\downarrow}\,\frac{
\mathsf{ac}{\downarrow}\,\dfrac{S\{R[a,a]\,[x/\tau]\}}{S\{\exists x R[a,a]\}}
}{S\{\exists x R\{a\}\}}\ .
$$

Weakening is separated as in the propositional case. When moved over $\mathsf{n}{\downarrow}$ and $\mathsf{n}{\uparrow}$, derivations of weakenings will contain weakenings on different atoms (with the same predicate symbol but differently instantiated):

$$
\mathsf{n}{\downarrow}\,\frac{
\mathsf{aw}{\uparrow}^n\,\dfrac{S'\{R'\{a\}\,[x/\tau]\}}{S'\{\exists x R'\{a\}\}}
}{S\{\exists x R\{\mathsf{t}\}\}}
\quad \rightsquigarrow \quad
\mathsf{aw}{\uparrow}^n\,\frac{
\mathsf{n}{\downarrow}\,\dfrac{S'\{R'\{a\}\,[x/\tau]\}}{S\{R\{\mathsf{t}\}\,[x/\tau]\}}
}{S\{\exists x R\{\mathsf{t}\}\}}
$$

$\square$

7.7 Decomposition and the Sequent Calculus

Decomposition, in a limited form, also exists in the sequent calculus: the separation of identity and cut as in Theorem 7.2.3, for example. The possibility of restricting weakening to the leaves of a sequent calculus proof is also a decomposition into a lower phase without weakening and an upper phase with weakening.

Clearly, weakening can not be restricted to the bottom (i.e. the root) of a sequent calculus proof, as deep inference allows us in Theorem 7.4.1. Just consider the valid sequent

$$\vdash (A \vee \bar{A}) \vee B \quad ,$$

for which there is no proof in GS1p with weakening restricted to the bottom.

Contraction can of course easily be restricted to the bottom of a sequent calculus proof in systems with additive context treatment, such as system G3cp, simply because in such systems contraction is admissible. However, this has nothing to do with decomposition, since contraction has just moved into the context management of the additive rules. Anyway, this trick does not work in the predicate case: moving to additive context treatment does not make contraction admissible.

The separation of contraction in Theorem 7.3.1 is such that in the upper phase no duplication of formulas takes place. To achieve this in sequent systems, one would have to restrict contraction to the bottom of a proof with multiplicative context treatment. This is impossible:

Proposition 7.7.1. *There is a valid sequent that has no proof in multiplicative* GS1p *in which all contractions are at the bottom.*

Proof. Consider the following sequent:

$$\vdash a \wedge a, \bar{a} \wedge \bar{a} \quad . \tag{7.1}$$

It suffices to show that, for any number of occurrences of the formulas $a \wedge a$ and $\bar{a} \wedge \bar{a}$, the sequent

$$\vdash a \wedge a, \ldots, a \wedge a, \bar{a} \wedge \bar{a}, \ldots, \bar{a} \wedge \bar{a} \tag{7.2}$$

is not provable in GS1p without contraction. Since the connective $\vee$ does not occur in this sequent, the only rules that can appear in contraction-free derivations with this endsequent are Ax, R∧ and RW. The only formulas that can appear in such derivations are $a \wedge a, \bar{a} \wedge \bar{a}, a$ and $\bar{a}$. Consequently, the only formulas that can appear in an axiom are the atoms a and $\bar{a}$. A leaf can thus be closed with an axiom only if it contains exactly two atoms (as opposed to two non-atomic formulas).

We prove by induction on the size of the derivation that each such derivation has a leaf which contains at most one atom. The base case is trivial: the sequent 7.2 contains no atom. For the inductive case, consider a derivation Δ. Remove a rule instance ρ from the top of Δ, to obtain a derivation Δ'. Let l be the leaf with the conclusion of ρ. By inductive hypothesis, Δ' has a leaf with at most one atom. Assume that this leaf is l, otherwise the inductive step is trivial. The rule instance ρ can not be an axiom, because there is at most one atom in l. If ρ is a weakening then the premise of ρ contains at most one atom. If ρ is an instance of R$\wedge$ then the only atom that may occur in the conclusion goes to one premise. The other premise contains at most one (i.e. exactly one) atom. $\square$

Similarly to the argument why contraction can not be restricted to atomic form in the sequent calculus (Proposition 4.3.1), the above works not only for system GS1p but for a wide range of sequent systems, as long as the R$\wedge$-rule is multiplicative.

7.8 Interpolation

In system SKS there are rules that, when going up in a derivation, introduce new atoms: the cut rule and the co-weakening rule. An interesting question to ask is whether one can restrict those rules such that they only introduce atoms that appear in the premise of the derivation. Cut elimination answers this question in the case of proofs, which are special derivations. If the premise of a derivation is the constant t, then cut and co-weakening are superfluous—no 'detours' are necessary. What about derivations that are not proofs? Further, in the case of derivations (as opposed to proofs) we have a symmetry between premise and conclusion. This suggests to symmetricise the above question: Can we restrict cut and co-weakening to atoms appearing in the premise and at the same time restrict identity and weakening to atoms appearing in the conclusion? Guglielmi stated this as a conjecture in [20].

In this section I will give a decomposition theorem for derivations in system SKS, which yields as immediate corollaries not only this conjecture (and thus cut elimination) but also (propositional) Craig interpolation [10]. The proof will use semantic means.

Definition 7.8.1. A formula is in *disjunctive normal form* if it is a disjunction of conjunctions of atoms. Given a formula R, a formula in disjunctive normal form that is semantically equivalent to R is denoted by $dnf(R)$. A formula is in *canonical disjunctive normal form* if it is in disjunctive normal form and 1) each conjunction occurs at most once and 2) each propositional variable occurring in the formula occurs exactly once in each conjunction, either negated or not negated, but not both. Given a formula R, a formula in canonical disjunctive normal form that is semantically equivalent to R will be denoted by $cdnf(R)$.

Example 7.8.2. The formula $(a, [b, c])$ is not in disjunctive normal form, but the semantically equivalent formula $[(a, b), (a, c)]$ is. However, it is not in canonical disjunctive normal form, while the semantically equivalent formula $[(a, b, c), (a, b, \bar{c}), (a, \bar{b}, c)]$ is.

Lemma 7.8.3. *For every derivation* $\begin{array}{c} P \\ \| \mathsf{SKS} \\ Q \end{array}$ *there is a derivation of the following shape:*

$$
\begin{array}{c} P \\ \Big\|_{\{c\uparrow,s\}} \\ dnf(P) \\ \Delta_1 \Big\| \\ cdnf(P) \\ \Big\|_{\{w\downarrow\}} \\ cdnf(Q) \\ \Delta_2 \Big\| \\ dnf(Q) \\ \Big\|_{\{c\downarrow,m\}} \\ Q \end{array}
\qquad with \qquad
\Delta_1 =
\begin{array}{c} dnf(P) \\ \Big\|_{\{aw\uparrow\}} \\ P_1 \\ \Big\|_{\{c\downarrow\}} \\ P_2 \\ \Big\|_{\{aw\uparrow,ai\uparrow\}} \\ P_3 \\ \Big\|_{\{c\uparrow,ai\downarrow,s\}} \\ cdnf(P) \end{array}
\qquad and \qquad
\Delta_2 =
\begin{array}{c} cdnf(Q) \\ \Big\|_{\{c\downarrow,aw\uparrow\}} \\ Q_3 \\ \Big\|_{\{w\downarrow\}} \\ Q_2 \\ \Big\|_{\{w\downarrow\}} \\ Q_1 \\ \Big\|_{\{ac\uparrow\}} \\ dnf(Q) \end{array}
\quad.
$$

Proof. A disjunctive normal form can be derived from P, and Q can be derived from one of its disjunctive normal forms, respectively, by using repeatedly the following derivations:

$$
\mathsf{s}^2\dfrac{\mathsf{c}\uparrow\dfrac{S(R, [T, U])}{S(R, R, [T, U])}}{S[(R, T), (R, U)]}
\qquad and \qquad
\mathsf{c}\downarrow\dfrac{\mathsf{m}\dfrac{S[(R, T), (R, U)]}{S[(R, R], [T, U])}}{S(R, [T, U])}
\quad.
$$

Since P implies Q, all successful valuations of P are successful on Q. A successful valuation of a formula (i.e. a line in the truth table that satisfies the formula) corresponds to a conjunction in a canonical disjunctive form of the formula. To derive $cdnf(Q)$ from $cdnf(P)$ it thus suffices to remove conjunctions from $cdnf(Q)$ which can be done by applying the $\mathsf{w}\downarrow$ rule.

What remains to be shown is the existence of Δ_1 and Δ_2, i.e. how to derive a canonical disjunctive normal form from a disjunctive normal form and vice versa. This is done in four steps, in Δ_1 seen top-down and in Δ_2 seen bottom-up:

1. In each conjunction, for each atom, all occurrences except for one are removed.

$$\mathsf{aw}{\uparrow}\, \frac{[S, (a_1, a_1, a_2, \ldots, a_n)]}{[S, (a_1, a_2, \ldots, a_n)]} \qquad \mathsf{ac}{\uparrow}\, \frac{[S, (a_1, a_2, \ldots, a_n)]}{[S, (a_1, a_1, a_2, \ldots, a_n)]}$$

2. For each conjunction, all occurrences of it except for one are removed.

$$\mathsf{c}{\downarrow}\, \frac{[S, (a_1, \ldots, a_n), (a_1, \ldots, a_n)]}{[S, (a_1, \ldots, a_n)]} \qquad \mathsf{w}{\downarrow}\, \frac{[S, (a_1, \ldots, a_n)]}{[S, (a_1, \ldots, a_n), (a_1, \ldots, a_n)]}$$

3. All unsatisfiable conjunctions, i.e. those that contain an atom together with its dual, are removed.

$$\mathsf{w}{\uparrow}\, \frac{[S, (a, \bar{a}, b_1, \ldots, b_n)]}{\mathsf{ai}{\uparrow}\, \dfrac{[S, (a, \bar{a})]}{S}} \qquad \mathsf{w}{\downarrow}\, \frac{S}{[S, (a, \bar{a}, b_1, \ldots, b_n)]}$$

4. For each propositional variable p that occurs in P or Q, and for each conjunction $(a_1, \ldots, a_n)$, in which p does not occur, remove this conjunction and replace it by $[(a_1, \ldots, a_n, p), (a_1, \ldots, a_n, \bar{p})]$. In Δ_1 we have

$$\mathsf{ai}{\downarrow}\, \frac{\mathsf{c}{\uparrow}\, \dfrac{[S, (a_1, \ldots, a_n)]}{[S, (a_1, \ldots, a_n, a_1, \ldots, a_n)]}}{\mathsf{s}^2\, \dfrac{[S, (a_1, \ldots, a_n, a_1, \ldots, a_n, [p, \bar{p}])]}{[S, (p, a_1, \ldots, a_n), (\bar{p}, a_1, \ldots, a_n)]}} \qquad ,$$

and in Δ_2 we have

$$\mathsf{c}{\downarrow}\, \frac{\mathsf{aw}{\uparrow}^2\, \dfrac{[S, (p, a_1, \ldots, a_n), (\bar{p}, a_1, \ldots, a_n)]}{[S, (a_1, \ldots, a_n), (a_1, \ldots, a_n)]}}{[S, (a_1, \ldots, a_n)]} \qquad .$$

$\square$

Using the lemma above, a derivation is separated into two phases: the top one, with rules that do not introduce new atoms going down, and the bottom one, with rules that do not introduce new atoms going up. Consequently, the formula in between contains only atoms that occur both in the premise and in the conclusion of the derivation and is thus an *interpolant*. This decomposition theorem can be seen as the symmetric closure of cut elimination: not only are cuts pushed up, but also their duals, identities, are pushed down.

Theorem 7.8.4 (Interpolation).

$$\text{For every derivation } \begin{array}{c} P \\ \| \text{SKS} \\ Q \end{array} \text{ there is a derivation } \begin{array}{c} P \\ \| \text{SKS} \backslash \{ai\downarrow, aw\downarrow\} \\ V \\ \| \text{SKS} \backslash \{ai\uparrow, aw\uparrow\} \\ Q \end{array}.$$

Proof. Decompose the given proof by the above lemma. The derivation obtained, with $V = P_3$, almost has the desired shape: the only problem are instances of $aw\uparrow$ below P_3. They can be permuted up using the transformations given in the proof of Theorem 7.4.3. These transformations do not introduce new instances of cut or identity. □

Corollary 7.8.5 (Craig Interpolation). *For all formulas P and Q, if P implies Q then there is a formula V such that P implies V, V implies Q and V contains only propositional variables that occur in both P and Q.*

Proof. The derivation from P to V which exists by the interpolation theorem does not introduce new propositional variables when seen top-down. Neither does the derivation from V to Q when seen bottom-up. □

Corollary 7.8.6 (Cut Elimination).

$$\text{For each proof } \begin{array}{c} \| \text{SKS} \\ Q \end{array} \text{ there is a proof } \begin{array}{c} \| \text{SKS} \backslash \{ai\uparrow\} \\ Q \end{array}.$$

Proof. Applying the interpolation theorem on a given proof of Q, we get a proof

$$\begin{array}{c} t \\ \| \text{SKS} \backslash \{ai\downarrow, aw\downarrow\} \\ V \\ \| \text{SKS} \backslash \{ai\uparrow, aw\uparrow\} \\ Q \end{array}$$

In the derivation below V there are no cuts. Assume that there is a cut in the derivation above V. Since there is no rule that, going up, could remove the atoms introduced by a cut, these atoms have to occur in the premise of the proof. But the premise of a proof is the constant t, so it cannot contain atoms. Thus the proof is cut-free.   □

Of course, Craig interpolation is easily obtained from cut elimination in the sequent calculus, so it comes as no surprise that a theorem that easily yields cut elimination also yields Craig interpolation with relative ease. However, what we have here is not just easy, it is immediate: cut elimination and Craig interpolation can immediately be read off of an easy-to-state normal form of derivations. So the given interpolation theorem is a simple generalisation of both: cut elimination as well as Craig interpolation.

Chapter Summary

We have seen that deep inference allows permutations that can not be observed in the sequent calculus, which leads to new normal forms for derivations. Cut and identity, weakening and co-weakening, as well as contraction and co-contraction can all be separated from the other rules. These normal forms hold both in propositional and in predicate logic. Except for the first, they are impossible in the sequent calculus.

One normal form, for propositional logic, is especially interesting since it generalises both cut elimination and Craig interpolation.

Chapter 8

Conclusion

We have seen deductive systems for classical propositional and predicate logic in the calculus of structures. They are related to sequent systems, but their rules apply *deep* inside formulas, and derivations enjoy a top-down *symmetry* which allows to dualise them. Just like sequent systems, they have a cut rule which is admissible, so they in principle admit a proof theory similar to sequent systems.

In contrast to sequent systems, they allow to reduce the cut, weakening and contraction to atomic form. This leads to *local* rules, i.e. rules that do not require the inspection of expressions of unbounded size. For propositional logic, I presented system SKS, which is local, i.e. contains only local rules. For predicate logic I presented system SKSq which is local except for the treatment of variables.

The reducibility of cut to atomic form together with deep inference also allows to obtain *finitely generating* systems, that is, systems in which each rule has only a finite choice of premises once the conclusion is given. This can be done in a very simple manner, cut elimination is not needed for obtaining finite choice.

Another consequence of the reducibility of cut to atomic form is a cut elimination procedure which is much simpler than those for sequent systems. The way in which proofs are substituted resembles normalisation in natural deduction, which suggests a computational interpretation in the proof normalisation as computation paradigm. Since the systems have an explicit admissible cut rule, they are in principle suitable for proof search as computation. So each of the systems seems to be a good candidate for developing both the proof search as well as the proof normalisation paradigm together in the same system.

The freedom in applying inference rules in the calculus of structures allows permutations that can not be observed in the sequent calculus. This leads to new normal forms for derivations, as shown in the decomposition theorems. One of these normal forms is especially interesting since it generalises both cut elimination and Craig interpolation.

These results show that deep inference and top-down symmetry allow for a more refined analysis of proofs than the sequent calculus. I think the development of a proof theory based on these two concepts is promising. I would therefore like to show in the following some open problems which, I believe, are worthy of further research.

Cut Elimination for Predicate Logic

A natural question is whether the cut elimination procedure for system SKS scales to more expressive cases, for example to predicate logic. At this point, the proof of cut admissibility for SKSq relies on cut admissibility in the sequent calculus. The cut elimination procedure presented in Chapter 6 does not appear to easily scale to system SKSq. The problem, which does not occur in shallow inference systems like sequent calculus or natural deduction, are existential quantifiers in the context of a cut which bind variables both in a and $\bar{a}$. The procedure easily extends to *closed* atomic cuts, where the cut formula is an atom prefixed by quantifiers that bind all its variables. The question then is how to reduce general cuts to closed atomic cuts. If this problem were solved, then the procedure would scale to predicate logic. Hopefully this will lead to a cut elimination procedure for predicate logic which is simpler than other cut elimination procedures, as happened for propositional logic.

Complexity of Cut Elimination

The procedure given in Chaper 6 involves a massive increase in the size of the proof, since for each cut the proof above it is duplicated. Another technique, splitting [22], can avoid much of this duplication by separating the context of the cut into two formulas and the proof above the cut into two proofs: one for each cut atom together with the corresponding part of the context. It will be interesting to study the complexity of such a procedure and how it relates to the one presented here and the ones for sequent systems.

Intuitionistic Logic

Which new proof-theoretical properties can be observed for intuitionistic logic by using deep inference and symmetry? Even though neither the De

Morgan laws nor the law of the contrapositive hold intuitionistically, there are systems for intuitionistic logic which enjoy the same symmetry as system SKS [5, 11]. This symmetry allows to reduce the cut to atomic form just like in SKS. The current question is whether there is a local system, i.e. whether contraction is reducible to atomic form, and whether the cut elimination procedure for SKS can be adapted to these systems.

Computational Interpretation of Cut Elimination

For proof normalisation as computation, natural questions to be considered are strong normalisation and confluence of the cut elimination procedure from Chapter 6—when imposing as little strategy as possible. Similarly to [30], a term calculus should be developed and its computational meaning be made precise. Intuitionistic logic is a more familiar setting for this, so the possibility of treating intuitionistic logic should be explored first.

Normal Forms

A decomposition theorem that has been proved for two other systems [24, 41] in the calculus of structures and led to cut elimination, is the separation of the *core* and the *non-core* fragment. So far, all the systems in the calculus of structures allow for an easy reduction of both cut and identity to atomic form by means of rules that can be obtained in a uniform way. Those rules are called the *core* fragment. In SKS, the core consists of one single rule: the switch. The core of SKSq, in addition to the switch rule, also contains the rules u↓ and u↑. All rules that are not in the core and are not identity or cut are called *non-core*. The main problem in separating core from non-core is separating switch from medial.

Conjecture (Separation core – non-core).

$$
\text{For every derivation } \begin{array}{c} T \\ \|\,\textsf{SKS} \\ R \end{array} \text{ there is a derivation } \begin{array}{c} T \\ \|\,non\text{-}core \\ T' \\ \|\,\{\textsf{ai}\!\downarrow\} \\ T'' \\ \|\,core \\ R'' \\ \|\,\{\textsf{ai}\!\uparrow\} \\ R' \\ \|\,non\text{-}core \\ R \end{array} \quad .
$$

A cut elimination procedure that is based on permuting up instances of the cut would be easy to obtain, could we rely on this conjecture. Then all the problematic rules that could stand in the way of the cut can be moved either below all the cuts or to the top of the proof, rendering them trivial, since their premise is the unit t. Cut elimination is thus an easy consequence of such a decomposition theorem. Straßburger suggests that decomposition theorems of this kind and cut elimination are closely connected [40].

Of course, one can imagine many different decomposition theorems, corresponding to different orders in which certain inference rules are applied. A decomposition theorem implies a normal form for derivations and thus a notion of equivalence of derivations. It is a well-known problem in proof theory that most formalisms distinguish between proofs that only differ by some trivial permutation of rule instances. The equivalence of proofs in formal systems has little to do with the intuitive notion of equivalence of proofs that mathematicians have. An interesting question is whether there is a decomposition such that two derivations have the same normal form exactly when the two derivations are intuitively equivalent. A starting point could be [23].

In addition to proving new decomposition theorems, new ways of proving existing decomposition theorems should be studied. The proof of the separation of contraction (Theorem 7.3.2), for instance, relies on the admissibility of the cut. It should be provable directly, i.e. without using cut admissibility, just by very natural permutations. The difficulty is in proving termination of the process of bouncing contractions up and down between cuts and identities, as happens in [42]. Another example is the interpolation theorem, which relies on Lemma 7.8.3, and thus on semantics. A syntactic procedure that just uses permutation of rules would be desirable, especially the bounds that it puts on the size of the interpolant.

Proof Search

The greater freedom in applying inference rules that is given by deep inference is a mixed blessing. On one hand it in principle allows to find shorter proofs than in the sequent calculus. On the other hand it implies a greater nondeterminism in proof search. It will be interesting to see whether it is possible to restrict this nondeterminism by finding a suitable notion of goal-driven proof like the notion of *uniform proofs* by Miller et al. [29]. Decomposition theorems restrict the choice of *which* inference rules to apply. A splitting theorem [22] restricts the choice of *where* to apply inference rules. In a goal-driven proof the application of certain inference rules could be restricted to the goal until there is no other choice besides applying them to the program. It is conceivable to develop various notions of goal-driven proof based on a combination of splitting and decomposition.

Relation to other Formalisms

The calculus of structures identifies structural and logical level as found in Gentzen's sequent calculus. Belnap's Display Calculus [2] can be seen as strengthening the structural level to match the logical level. It will be interesting to see the connection between the two. It would be especially interesting to see whether easily checkable syntactic criteria for cut elimination, as they exist for the Display Calculus, can be found for the calculus of structures.

The relation of the presented systems to the connection method [45] is another interesting subject. The finer granularity of inference rules in the calculus of structures with respect to the sequent calculus would presumably allow for a closer correspondence between derivations in the calculus of structures and connection proofs than between sequent calculus derivations and connection proofs. The connection method also has a certain symmetry which might help in establishing a correspondence.

Bibliography

[1] Alan Ross Anderson, Nuel D. Belnap, and J. Michael Dunn. *Entailment: The Logic of Relevance and Necessity*, volume 1. Princeton University Press, 1975.

[2] Nuel D. Belnap, Jr. Display logic. *Journal of Philosophical Logic*, 11:375–417, 1982.

[3] Kai Brünnler. Locality for classical logic. Technical Report WV-02-15, Dresden University of Technology, 2002. Available at http://www.wv.inf.tu-dresden.de/~kai/LocalityClassical.pdf.

[4] Kai Brünnler. Atomic cut elimination for classical logic. In M. Baaz and J. A. Makowsky, editors, *CSL 2003*, volume 2803 of *Lecture Notes in Computer Science*, pages 86–97. Springer-Verlag, 2003.

[5] Kai Brünnler. Minimal logic in the calculus of structures, 2003. Note. On the web at: http://www.ki.inf.tu-dresden.de/~kai/minimal.html.

[6] Kai Brünnler. Two restrictions on contraction. *Logic Journal of the IGPL*, 11(5):525–529, 2003.

[7] Kai Brünnler and Alessio Guglielmi. A first order system with finite choice of premises. In Hendricks et al., editor, *First-Order Logic Revisited*. Logos Verlag, 2004. To appear.

[8] Kai Brünnler and Alwen Fernanto Tiu. A local system for classical logic. In R. Nieuwenhuis and A. Voronkov, editors, *LPAR 2001*, volume 2250 of *Lecture Notes in Artificial Intelligence*, pages 347–361. Springer-Verlag, 2001.

[9] Paola Bruscoli. A purely logical account of sequentiality in proof search. In Peter J. Stuckey, editor, *Logic Programming, 18th International Conference*, volume 2401 of *Lecture Notes in Artificial Intelligence*, pages 302–316. Springer-Verlag, 2002.

[10] W. Craig. Linear reasoning. A new form of the Herbrand-Gentzen theorem. *Journal of Symbolic Logic*, 22:250–268, 1957.

[11] Philippe de Groote. Personal communication, 2001.

[12] Pietro Di Gianantonio. Structures in cyclic linear logic. Technical report, Università di Udine, 2003.

[13] Kosta Došen and Zoran Petrić. Bicartesian coherence. *Studia Logica*, 71(3):331–353, 2002.

[14] Roy Dyckhoff. Contraction-free sequent calculi for intuitionistic logic. *The Journal of Symbolic Logic*, (57):795–807, 1992.

[15] Jean Gallier. Constructive logics. Part I: A tutorial on proof systems and typed λ-calculi. *Theoretical Computer Science*, 110:249–339, 1993.

[16] Gerhard Gentzen. Investigations into logical deduction. In M. E. Szabo, editor, *The Collected Papers of Gerhard Gentzen*, pages 68–131. North-Holland Publishing Co., Amsterdam, 1969.

[17] Jean-Yves Girard. *Interprétation fonctionelle et élimination des coupures de l'arithmétique d'ordre supérieur*. PhD thesis, Université Paris VII, 1972.

[18] Jean-Yves Girard. Linear logic. *Theoretical Computer Science*, 50:1–102, 1987.

[19] Alessio Guglielmi. The calculus of structures website. Available from http://www.ki.inf.tu-dresden.de/~guglielm/Research/.

[20] Alessio Guglielmi. Goodness, perfection and miracles. On the web at: http://www.ki.inf.tu-dresden.de/~guglielm/Research/Notes/AG1.pdf, 2002.

[21] Alessio Guglielmi. Recipe. Manuscript. http://www.wv.inf.tu-dresden.de/~guglielm/Research/Notes/AG2.pdf, 2002.

[22] Alessio Guglielmi. A system of interaction and structure. Technical Report WV-02-10, Technische Universität Dresden, 2002. Available at http://www.wv.inf.tu-dresden.de/~guglielm/Research/Gug/Gug.pdf.

[23] Alessio Guglielmi. Normalisation without cut elimination. Manuscript. Available on the web at http://www.wv.inf.tu-dresden.de/~guglielm/Research/Notes/AG2.pdf, 2003.

[24] Alessio Guglielmi and Lutz Straßburger. Non-commutativity and MELL in the calculus of structures. In L. Fribourg, editor, *CSL 2001*, volume 2142 of *Lecture Notes in Computer Science*, pages 54–68. Springer-Verlag, 2001.

[25] Alessio Guglielmi and Lutz Straßburger. A non-commutative extension of MELL. In Matthias Baaz and Andrei Voronkov, editors, *Logic for Programming, Artificial Intelligence, and Reasoning, LPAR 2002*, volume 2514 of *LNAI*, pages 231–246. Springer-Verlag, 2002.

[26] J. Herbrand. *Recherches sur la théorie de la démonstration*. PhD thesis, Université de Paris, 1930.

[27] J. W. Lloyd. *Foundations of Logic Programming*. Springer-Verlag, 1993.

[28] Dale Miller. lambda Prolog: An introduction to the language and its logic. Draft of book in circulation since Summer 1995.

[29] Dale Miller, Gopalan Nadathur, Frank Pfenning, and Andre Scedrov. Uniform proofs as a foundation for logic programming. *Annals of Pure and Applied Logic*, 51:125–157, 1991.

[30] M. Parigot. $\lambda\mu$-calculus: an algorithmic interpretation of classical natural deduction. In *LPAR 1992*, volume 624 of *Lecture Notes in Computer Science*, pages 190–201. Springer-Verlag, 1992.

[31] D.J. Pym. *The Semantics and Proof Theory of the Logic of Bunched Implications*, volume 26 of *Applied Logic Series*. Kluwer Academic Publishers, 2002.

[32] Greg Restall. *An Introduction to Substructural Logics*. Routledge, 2000.

[33] John Alan Robinson. A machine-oriented logic based on the resolution principle. *Journal of the ACM*, 12:23–41, 1965.

[34] Kurt Schütte. Schlusssweisen-Kalküle der Prädikatenlogik. *Mathematische Annalen*, 122:47–65, 1950.

[35] Kurt Schütte. *Beweistheorie*. Springer-Verlag, 1960.

[36] Kurt Schütte. *Proof Theory*. Springer-Verlag, 1977.

[37] Raymond M. Smullyan. Analytic cut. *The Journal of Symbolic Logic*, 33:560–564, 1968.

[38] Raymond M. Smullyan. *First-Order Logic*. Springer-Verlag, Berlin, 1968.

[39] Charles Stewart and Phiniki Stouppa. A systematic proof theory for several modal logics. Technical Report WV-03-08, Technische Universität Dresden, 2003. Submitted.

[40] Lutz Straßburger. *Linear Logic and Noncommutativity in the Calculus of Structures*. PhD thesis, Technische Universität Dresden, 2003.

[41] Lutz Straßburger. A local system for linear logic. In Matthias Baaz and Andrei Voronkov, editors, *Logic for Programming, Artificial Intelligence, and Reasoning, LPAR 2002*, volume 2514 of *LNAI*, pages 388–402. Springer-Verlag, 2002.

[42] Lutz Straßburger. MELL in the Calculus of Structures. *Theoretical Computer Science*, 309(1–3):213–285, 2003.

[43] Alwen Fernanto Tiu. Properties of a Logical System in the Calculus of Structures. Master's thesis, Technische Universität Dresden, 2001.

[44] Anne Sjerp Troelstra and Helmut Schwichtenberg. *Basic Proof Theory*. Cambridge University Press, 1996.

[45] W. Bibel. On matrices with connections. *Journal of the Association for Computing Machinery*, 28(4):633–645, 1981.

[46] Philip Wadler. 19'th century logic and 21'st century programming languages. *Dr Dobbs*, December 2000.

Index